On the Rocks

On the Rocks

Earth Science for Everyone

John S. Dickey, Jr.
Trinity University
San Antonio, Texas

John Wiley & Sons, Inc.
New York • Chichester • Brisbane • Toronto • Singapore

This text is printed on acid-free paper.

Library of Congress Cataloguing in Publication Data:

Dickey, John S.
On the rocks : earth science for everyone / John S. Dickey.
p. cm.
Includes bibliographical references (p. –) and index.
ISBN 0-471-13234-9 (cloth : alk. paper)
1. Earth sciences—Popular works. I. Title.
QE31.D485 1996
550—dc20 95-46167

Printed in the United States of America

10 9 8 7 6 5 4 3 2 1

Dedicated to
my mother, Christina Gillespie Dickey,
my wife, Lynn McMath Dickey, and
my son, Nathaniel Hudson Dickey.

Contents

Preface

Sir Isaac Newton, the greatest scientist of all time, has given us a delightful description of scientific inquiry:

> . . . I seem to have been only a boy playing on the sea-shore, and diverting myself in now and then finding a smoother pebble or a prettier shell than ordinary, whilst the great ocean of truth lay all undiscovered before me.

The great ocean of undiscovered truth is still there, and so are the pebbles and shells, and although few of us can view them with anything like Newton's perception, any one of us can share his pleasure. This little book is intended to encourage and promote that sharing, especially among those who, for one reason or another, have not had an opportunity to study natural science.

The second half of the twentieth century has been a wonderful time, perhaps the best time ever, to do science. In the United States, which has led the world in science education and research productivity during this period, there has been steady growth in support for science and scientists, and that support has resulted in breathtaking advances in scientific knowledge and technology. To have been among the fortunate men and women who were born at the right time and educated in the right ways to participate in this extraordinary episode of human achievement was for me an enormous privilege. Science and scientists have filled my life, from childhood on, with joy and excitement.

This little book attempts to share some of that excitement with others who have not had the pleasure of personal involvement with science and scientists and to support the point of view, which is not always shared by nonscientists, that, in addition to being a magnificent intellectual adventure, the pursuit of science contributes to the well-being of humanity. This book is bullish on science. It confronts science phobes and mavens of uncertainty by presenting science as "the most effective technique yet devised for exploring the universe with the human mind." Without minimizing the threats of environmental degradation to human survival, it advocates an optimistic approach to our world and suggests that science "is also *potentially* the most effective technique for bettering the lives of all who live on Earth."

My subject is the science that used to be called geology but has grown and diversified into the earth and planetary sciences, which engage the disciplines of geochemistry and geophysics as well as classical geology with all of the condensed objects in our solar system. The book makes no attempt to provide a comprehensive review of earth and planetary science and is addressed, without hesitation or apology, at the dilettante rather than the serious student. Technical discussions are leavened by frequent vignettes describing individuals and events that figure prominently in the history of the science. I hope the result is a book that will be read with pleasure by a general audience.

San Antonio, Texas
September 1995

Acknowledgments

I would like to thank my students, friends, and colleagues who have helped me in this work. I am especially indebted to the following for their advice, criticism, encouragement, and technical assistance: Walter Alvarez, Craig F. Bohren, John A. Burke, Robert L. Freed, Stanley R. Hart, David H. Hough, Frank L. Kersnowski, Kenneth Kyle, Glenn C. Kroeger, Henry O. A. Meyer, Henry T. Mullins, Cathryn R. Newton, Frank Press, Stephen H. Richardson, Joseph E. Robinson, Donald I. Siegel, Raymond Siever, Norman Sherry, Richard Swope, Mahbub Uddin, and Hatten S. Yoder, Jr. Without their help the text would still be blemished with numerous errors of fact and interpretation. For those errors that may remain I take full responsibility.

Foreword

Science is the mark of our century; it is our cathedral, our pyramid, our great adventure. It would be a pity to have lived at this time and be ignorant of the nature of science. This is a book about earth and planetary science, the scientists who study Earth and the planets, and the nature of science in general. Why you should understand your home and its environment is self-evident, but you may not know why you should understand science. You should understand science because science is demonstrably the most effective technique yet devised for exploring the universe with the human mind.

Science is a way of thinking, but there are lots of ways of thinking. Why is science so powerful? Because science is built by quantitative measure and reproducible experiment or observation. Reproducibility is the essential requirement of a scientific fact. In this respect science differs profoundly from authoritarian ways of thinking. As an authority I might tell you that the Earth is flat. As a scientist you would insist that I provide a quantitative demonstration of its flatness, and you would check my work by making measurements yourself.

Is this simply rhetoric? Not at all. Call two or more friends who live at least five hundred miles north or south of you in the same time zone. Arrange for everyone to go out on a clear night and simultaneously measure the angle between the north star and the horizon. Compare your measurements. Is the Earth flat or round?

Every scientific fact must be susceptible to testing. Scientists, like other people, have lots of ideas. They covet only those that can be tested. Science progresses by more than one path, but the most common appears to be:

Observation
Idea
Test
Failure
Idea or Observation
Test etc.

The failure step is crucial. That is the essential difference between dogmatic and scientific thought. The dogmatist will not admit the possibility of failure: the dogma is true. The scientist presumes the possibility of failure: the idea is probably flawed. Scientists learn more from failure than from non-failure and, therefore, push ideas until they fail.

* * *

> Galileo Galilei (1564–1642) on reading the book of Nature: Philosophy is written in this grand book—I mean the universe—which stands continually open to our gaze, but it cannot be understood unless one first learns to comprehend the language and interpret the characters in which it is written. It is written in the language of mathematics, and its characters are triangles, circles, and other geometrical figures, without which it is humanly impossible to understand a single word of it; without these, one is wandering about in a dark labyrinth.[1]

Since the days of Pythagoras (if not before), philosophers have wondered why numbers provide the most accurate descriptions of Nature. The Pythagoreans imbued numbers with transcendent significance. According to Aristotle, "they supposed the elements of numbers to be the elements of all things, and the whole heaven to be a musical scale and a number."[2]

One need not accept the objective reality of numbers to realize that something special is going on. In the words of the physicist, Eugene Wigner: "It is difficult to avoid the impression that a miracle confronts us here . . . The miracle of the appropriateness of the language of mathematics for the formulation of the laws of physics is a wonderful gift which we neither understand nor deserve."[3]

Mathematics exercises sovereignty over science and technology, but not necessarily priority. Sometimes mathematics precedes and enables progress, as the invention of binary arithmetic preceded digital computers by five hundred years; and sometimes mathematics follows and clarifies, as the calculus followed and clarified Kepler's laws of planetary motion. Today we are witnessing another application of new mathematics to science with the development of the Theory of Chaos and fractal geometry.

* * *

Chaos and the Laws of Fractals

Many scientists know what to do with mathematics.
Hardly anyone—dare I suggest no one?
Knows what to do with mathematicians.

Starry-eyed, except when tired,
From staying out too late with their thoughts,
Mathematicians have mystery,
Gambits never shown,
Except, perhaps, in the twinkling of an eye.

They sit too close to the seat of secrets,
Not for us to know,
Good and evil, life itself, or market trends.

Like painters and poets, mathematicians make patterns.
Painters make patterns with spots of paint.
Poets make patterns with words.
Mathematicians make patterns with ideas.
That's the difference.

Although painters and poets would deny it,
Mathematicians provide more accurate descriptions.
No one knows why.

Perhaps, as Plato believed,
Mathematical objects, not just their symbols,
Exist.

The curves, the polyhedra and the numbers,
Like rocks, exist somewhere.

The reality of some mathematics
Is undeniable.
Primes *are* primes, damnit.

Then someone throws a curve,
Like infinitessimals or the set of all sets,
Reality wavers,
Mortal minds are momentarily scrambled.
That's for mathematicians to worry about.

Scientists look for patterns.
Measure, test and define
Inequalities.

Demonstrably
The most effective technique yet devised
For exploring the universe with the human mind,
Science is written
In the language of mathematics.

Right again, Galileo!
But, in practice, not all sciences are equally mathematical.
Some are merely statistical.

The mathematical bases of astronomy and physics
Are more established
Than those of psychology,
Or biology,
Or even chemistry.

Is it natural, or merely expedient,
To leave mathematics at the laboratory door
When living creatures come to call?

Nature is oblivious.
The fault is ours alone.
Mathematics defines the cosmos,
From quarks to quasars,
From headaches to eternal love.

The stars
Drew mathematics and science
Together.

Until the seventeenth century
Astronomers related points of light only by angles
In an imperturbable universe.
Then Tycho saw the supernova of 1572,
Shattering the crystalline spheres of doomed philosophies.

He was only walking home for supper,
On a clear November evening.
He hadn't aimed to rock the boat.

The light of that new star,
Fixed in its station
By numbers read from Tycho's cross staff,
By angles drawn with Cassiopeia,
Was frozen still.

Too still for Aristotle's sublunary sphere.
That wise old man, his jig was up.
Cosmic change and evolution had appeared on the horizon.

Physics followed
In hot pursuit
Galileo, Newton, Faraday and Rutherford,
Defining Nature in pixels
Called experiments.

Going from points of light
To form and motion
Required some serious smoothing.

A chicken farmer dialed a wrong number at the local college and reached
The Physics Department rather than the Agricultural Extension Agent.
"My chickens won't lay. Can you help?" he asked.
Loath to admit ignorance, the physicist told him "call back in a week."
Which he did.

"We've found your problem," said the physicist.
"But we've had to make some simplifying assumptions.
Given a spherical chicken . . ."

Now we are told
New mathematics may facilitate
The mathematization of life sciences and other domains
Of naturally and necessarily complex forms.
Praises be and thanks!

To Pythagoras, Euclid, Euler, Leibniz, Gauss *et al.*
Eternal praises never ending,
Yet

In the late evening of our scientific century
Before the dawn, we ask
What's new?
Our friends reply,
"Try chaos and observe the laws of fractals."

* * *

True understanding of natural phenomena begins with comprehension of scale, whether one is talking of time, space, mass or energy. Scale determines which of the fundamental natural forces will dominate any phenomenon. Form *and* function, therefore, follow scale. The greatest intellectual challenge in science is to comprehend phenomena at scales that differ momentously from the scales of human experience. That occasionally we can meet this challenge is due to the astonishing power of numbers.

As numbers grow very large or very small they become more difficult to manipulate; therefore scale is often expressed in terms of powers of ten, otherwise referred to as orders of magnitude. These powers of ten are easily stated in exponential form. Ten to the fifth, or 10^5, for example, means 10 times 10 times 10 times 10 times 10, or 100,000. Although exponential notation is convenient, it tends to obscure the absolute differences between orders of magnitude. For example, the difference between 10^1 and 10^2 is only 90 (100 – 10), but the difference between 10^6 and 10^7 is 9,000,000 (10,000,000 – 1,000,000)!

Given the amazing effectiveness of numbers to describe the universe and of science as a way of thinking, are there inevitable limits to our comprehension? Some say yes,

> J.B.S. Haldane: . . . the universe is not only queerer than we suppose, but queerer than we can suppose.[4]

but if such limits exist they lie so far beyond our current understanding that we could not possibly recognize them.

Because it operates on the edges of understanding, science is inherently disruptive. Major advances are likely to be unsettling. The Copernican Revolution and Tycho Brahe's discoveries of cosmic change jostled sixteenth-century Europe; Darwin and Wallace upset the Victorians; now many physicists regard the nature, origin, and destiny of the universe as knowable; and then . . . But let Stephen Hawking speak for himself:

> Then we shall all, philosophers, scientists, and just ordinary people, be able to take part in the discussion of the question of why it is that we and the universe exist. If we find the answer to that, it would be the ultimate triumph of human reason—for then we would know the mind of God.*

If that presumptuous suggestion doesn't startle you, as we say in Texas, your wood's wet.

One

Gathering Stardust

Geology is an historical science. Geologists study the behavior of Earth and other planets as a function of time. As a geology student I expected to acquire a profound comprehension of time, and my professors seemed to deal confidently with millions or even billions of years. Perhaps they did comprehend these stupendous periods; perhaps some of my colleagues do now. I do not. I know most of the numbers (Earth is 4,550,000,000, or 4.55×10^9, years old, etc.), but I am still bewildered by the enormity of a million years, and I am equally mystified by very short time spans, like the femtosecond (10^{-15} sec) periods of nuclear reactions.

Am I alone? No. Most people have difficulty grasping time spans that differ markedly from human experience. Beyond the human time scale, which is from about one second to one century, my comprehension extends to ten millenia on the long side and a millisecond on the short side. Within this range I have a sense of temporal perspective, but for longer or shorter periods my senses are fuzzy; I lack perspective.

Yardsticks may help to overcome this human limitation. Museums often represent geologic time by a long stripe along a wall or up a spiral stairway. I will do the same:

____L_____________________________________H____D!

If the length of this line represents Earth history from the beginning on the left to the present on the right, then life may have appeared about at L, hard-shelled creatures at H, dinosaurs at D, and we humans at somewhere within the dot on the exclamation point. Alternatively, relative proportions of time periods may be understood by shrinking the entire period down to one year. If we do this to the age of the Earth, we find that the first hard-shelled creatures do not appear until Thanksgiving; the dinosaurs appear in mid-December and are wiped out on Christmas Eve; humans appear on New Year's Eve; and recorded history is accorded the last few seconds of the year.

Perhaps these tricks help, perhaps not. When all is said and done, the best approach to geologic time may be to let it wash over you like ocean waves—overwhelming, continuous, and utterly impossible to grasp.

Ideas about the formation of the Earth and our place in the universe often begin with star gazing. The universe defies comprehension, but that doesn't stop us. We ask more, and we learn more. This is a particularly exciting time in cosmology (the study of the universe), for we are blessed to be the first humans to study the universe from beyond the veil of Earth's atmosphere.

Gaze into the night sky. The deeper you look, the farther back you go in time. Moonlight reaches us 1.3 seconds after it leaves the Moon. Although you are seeing the Moon as it was 1.3 seconds ago, the scene is coming to you "live." Almost all of the starlight that falls on you left our galactic neighbors tens of thousands of years ago. Our galaxy, the Milky Way, is eighty thousand light years across. If you can make out the hazy spot that is M31, the Great Spiral Galaxy in the constellation Andromeda, the light you see left that galaxy when our hominid ancestors first began using stone tools on the savannas of subtropical Africa. M31 is two and a half million light years away, which means that its light took two and a half million years to reach us. Telescopes can carry you back still farther into deep space and time, back millions, even billions of years, long before our solar system existed, perhaps even to the beginning of time.

That there was a beginning of time seems likely. Consider the riddle known as *Olber's Paradox*: If the universe is infinitely old and filled with nothing but countless stars, then why is the night sky dark? If there are stars in every direction and enough time to warm every cloud of interstellar gas, why isn't the night sky aglow? Recent observations from space as well as Earth indicate that the universe of galaxies and other observable phenomena is not infinitely old but rather began with a big bang ten to twenty billion years ago. The galaxies and the rest of the observable universe have apparently been flying apart ever since, and in the eloquent words of Chet Ramo: "Stars glitter on a black cloth of night because the universe had a beginning."[6]

In addition to moving away from each other, galaxies also rotate. In our Milky Way the Sun orbits the galactic center at about 250 kilometers a second. One revolution takes 225 million years, so our Sun, which is about 5 billion years old, has made about 22 revolutions.

The centrifugal force associated with the rotations of the galaxies keeps the stars from falling into the galactic center. In fact, recent measurements indicate that the galaxies are spinning so fast that if their total mass were represented solely by the luminous stars, they would fly apart! Since they don't, there must be a lot more mass holding the galaxies together. According to Vera Rubin, an astronomer at the Carnegie Institution of Washington, "as much as 90 percent of the mass of the universe is non-luminous and is clumped, halo-like, around individual galaxies."[7] And though the nature of this mysterious dark matter is not known, its importance cannot be overstated. The visible portions of the cosmos are the basis for most of our theories about the origins of the cosmos and its components, including this planet. Until we know more about the invisible portions, our cosmology must be seriously deficient, but at the local level—the level of our solar system—we may know enough to answer some questions.

Galaxies are concentrations of billions of stars. Ours is a spiral, shaped like a disk, that is 80,000 light years across and up to 15,000 light years thick.* Our known planetary system, out to Pluto, which is 6 billion kilometers from the Sun, is approximately 11 light hours across. Earth is 8.3 light minutes from the Sun. The diameter of the galaxy is nearly eight orders of magnitude greater than the diameter

---

*Light travels 9.5×10^{12} kilometers per year, so the galaxy is

$$9.5\times10^{12}\ \text{km/yr} \times 8\times10^{4}\ \text{yr} = 7.6\times10^{17}\ \text{kilometers across.}$$

of our planetary system. For comparison, the diameter of Earth (12,742 kilometers) is eight orders of magnitude greater than the diameter of a large grapefruit (12.7 centimeters).

Theories on the origin of our solar system must explain several well-established facts. First is the consistency of motion. The planets are not flying about every which way. They all revolve around the Sun in the same direction and in nearly the same plane, and most moons revolve around their planets in this same direction. All the planets, except Venus, Uranus and Pluto, spin on more or less vertical axes (relative to the orbital plane) in the same direction as they revolve around the Sun. The Sun also spins, slowly, around a nearly vertical axis in the same direction as the revolving planets. This consistency of motion strongly implies that the Sun and its satellites originated by some fairly simple, continuous process.

Second, the distance of each planet from the Sun is not random. From Mercury to Uranus, each planet is roughly twice as far out as its inner neighbor. This relationship is approximated by the numerical series known as the *Titius-Bode Law*: Write the series of numbers 0, 3, 6, 12, 24, 48, 96, 192, 384, 768, and add 4 to each number in the series. Divide by 10 and compare the result to the distance of each planet from the Sun expressed in terms of the so-called astronomical unit (AU), Sun, which is defined as Earth's distance from the Sun.*

Titius-Bode Series	*Planet*	*Distance from the Sun*
0.4	Mercury	0.3871 AU
0.7	Venus	0.7233 AU
1.0	Earth	1.0000 AU
1.6	Mars	1.5237 AU
2.8	Ceres	2.7673 AU
5.2	Jupiter	5.2028 AU
10.0	Saturn	9.5388 AU
19.6	Uranus	19.1820 AU
38.8	Neptune	30.0577 AU
77.2	Pluto	39.5177 AU

*One astronomical unit (AU) equals 149,597,870 kilometers, which is about 93 million miles.

A German astronomer, Johann Daniel Titius, mentioned this empirical relationship in a footnote in 1766, and another German astronomer, Johann Elert Bode, publicized it in 1772. At that time the known solar system extended only to Saturn, and the asteroid belt, a ring of rocks and miniature planets (planetesimals) between Mars and Jupiter, was unknown. Uranus, discovered in 1781, gave additional credence to the Titius-Bode relationship, so Bode urged astronomers to search for a planet at 2.8 AU, between Mars and Jupiter. On the first night of the nineteenth century, January 1, 1801, Giuseppe Piazzi, Royal Astronomer and Director of the Royal Observatory at Palermo, discovered Ceres, the largest of the asteroids. Karl Friedrich Gauss (1777–1855), the great mathematician, calculated Ceres' orbit to be 2.79 AU from the Sun. A year later the second largest asteroid, Pallas, was discovered by Wilhelm Olbers, the same person whose famous paradox, quoted earlier, brought profound meaning to the darkness of night. There are now roughly five thousand recognized asteroids. Subsequent discoveries of Neptune in 1846 and Pluto in 1930, at 30.0 and 39.5 AU, respectively, invalidated the Titius-Bode relationship as a natural law, but they did not erase the evidence of some relationship between orbital radius and planet formation.

During the latter half of the eighteenth century, German philosopher Immanuel Kant (1724–1804) and French astronomer/mathematician Pierre Simon de Laplace (1749–1827) proposed the nebular hypothesis, in which the planets condensed from a cooling cloud of gas revolving around a primitive Sun. As the cloud cooled it contracted, and as it contracted, it revolved faster. With increasing velocity, the matter concentrated in rings and, later, in planetary bodies.

During the nineteenth century concern over the distribution of mass and angular momentum in the solar system weakened the nebular hypothesis, and it was replaced for several decades by hypotheses involving collisions or close encounters of the Sun with one or more stars. These theories, which were proposed by such great figures in astronomy and physics as James Jeans and Harold Jeffreys, eventually encountered their own difficulties, some more crippling than the problem of angular momentum.

Angular momentum is the mathematical product of a turning body's mass, times its velocity, times its distance out from the center of rotation:

$$\text{Angular momentum} = \text{mass} \times \text{velocity} \times \text{distance from axis of rotation}$$

Every object in the solar system has angular momentum, but although the Sun has 99.9 percent of the mass of the solar system, it has less than 1 percent of the system's angular momentum because it spins so slowly, rotating only once every 25 to 27 days. Galileo was the first to know this. In 1610 he watched sunspots move day by day, from west to east, across the face of the Sun, and he realized that this movement was due to the Sun's rotation. The Sun's angular momentum (1.5×10^{41} kg*m^2/sec) is two orders of magnitude less than the angular momentum of all the planets put together (3.15×10^{43} kg*m^2/sec), and this has been a big problem for the nebular hypothesis because, by natural law, angular momentum must be conserved.

Take a large stone and hold it in front of you. Start turning rapidly around and turn pull the stone in close to your body. As you do you will feel a strong force making you turn faster. This is conserving your angular momentum. Your increased velocity is making up for the decreased distance between the stone and your axis of rotation. According to the nebular hypothesis, the Sun and the planets were formed by the condensation and contraction of an enormous, rotating cloud. The Sun, however, with almost all of the mass of the solar system, turns too slowly on its axis to have been formed simply by contraction of a rotating cloud of gas. It is as though the gathering, spinning Sun had brought a huge stone closer to its body and yet had not spun faster.

The angular momentum problem was almost fatal for the nebular hypothesis, but the hypothesis has returned to favor in recent decades because mechanisms have been found to explain angular momentum and the distribution of mass. In fact, it is now believed that angular momentum must be transferred outward into a dust cloud if star formation is to occur. The mechanism for this transfer may involve interaction of the star's magnetic field with the cloud and frictional forces within the cloud.

* * *

Chemical constraints upon theories of the origin of the solar system came later than physical constraints because, until the nineteenth century, the chemistry of the cosmos was unknown and thought to be unknowable. The stars and planets were points of light, far beyond the reach of chemists. Then spectroscopy provided a way to determine the elemental compositions of luminous, distant objects.

In 1666, when Isaac Newton (1642–1727) spread white light into its spectral bands upon his chamber wall with a glass prism, the "phænomena of colours," as he called it, was already well known, but no one had yet recognized the composite nature of white light. What Newton discovered was that white light was a mixture of different colors, from red to violet, that were physically separable by their different velocities through transparent media such as glass. He discovered, too, that the colors of the separated rays could not be changed. White light could be created by recombining the colored rays, but the color of any one of the rays could not be changed by shining it through a colored filter, reflecting it off a colored mirror, or bending it with a prism.

The next significant step came in 1814, when a German optician, Joseph von Fraunhofer, observed the solar spectrum through a telescope and saw "an almost countless number of strong and weak vertical lines, which are, however, darker than the rest of the color-image; some appeared to be almost perfectly black."[8] Fraunhofer was seeing the absorption of specific wavelengths (colors) of light by atoms in the solar and terrestrial atmospheres.

Atoms can absorb light at specific wavelengths, causing the dark spectral lines seen by Fraunhofer, and they can emit light at the same specific wave lengths, creating bright spectral lines. Sodium, for example, absorbs and emits light at two wavelengths that are in the yellow-orange range of the spectrum. This causes the yellow-orange flash of light whenever something salty spills into a flame. The precise specificity of atomic emissions and absorptions is the basis of the *quantum theory*, which declares that energy is transferred only in specific amounts, or quanta. It is also the basis for chemical analysis by spectroscopy because the intensities of the emissions or absorptions are proportional to the concentrations of the various atoms in the material that is emitting or absorbing light.

In 1860 Robert Bunsen (immortalized as the inventor of a common laboratory gas burner) and Gustav Kirchhoff described the method of spectroscopic chemical analysis. Kirchhoff applied the method to the solar atmosphere, where he confirmed the presence of iron, calcium, magnesium, sodium, nickel, and chromium. Over the years since then spectral analysis has become routine and exceedingly versatile, using different forms of electromagnetic radiation, from X rays through visible light to the invisible infrared, and analyzing the elemental compositions of luminous objects from distant stars to household dust. The chemistry of the luminous universe, including the sunlit surfaces and atmospheres of planets, has become known.

The luminous universe analyzed by spectroscopy is an ocean of hydrogen with a dash of helium and minute traces of everything else. Is this also true of the mysterious dark matter that we can not see or otherwise detect yet suspect to be much more abundant than visible ordinary matter? No. We do not know what that invisible matter is. We know only that it does not behave as ordinary matter. We do, however, know the distribution of ordinary matter in our solar system.

* * *

There are terrestrial planets and giant planets. The terrestrial planets are composed mostly of iron (Fe), silicon (Si), oxygen (O), and magnesium (Mg); the giant planets more closely resemble the Sun because they are composed primarily of hydrogen (H), helium (He), oxygen (O), and carbon (C).

Because almost all of the mass of the present solar system is in the Sun, we assume that the material that formed the planets was like the Sun in chemical composition. The relative proportions of chemical elements heavier than nitrogen are the same in many meteorites as they are in the Sun, and for this reason we refer to these meteorites as *undifferentiated*. Spectral studies of the Sun and analyses of undifferentiated meteorites reveal the overall composition of the solar system. The proportions of the elements present, relative to one atom of silicon, are as follows:[9]

Hydrogen	27,200.0
Helium	2,180.0
Oxygen	20.1
Carbon	12.1
Neon	3.76
Nitrogen	2.48
Magnesium	1.07
Silicon	1.00
Iron	0.90
Sulfur	0.515
All other (82) elements	<1.00 Total

Stars and planetary systems like ours are believed to orginate in dense molecular clouds that exist between stars. Hydrogen and helium are the most abundant substances in these clouds, which are 99 percent gas, but more than eighty other substances (carbon monoxide, water, ammonia, and many compounds of carbon, hydrogen and oxygen) have also been identified. One percent of the interstellar clouds is dust. What are the dust grains made of? Silicates (compounds containing silicon and oxygen) and silicon carbide are known to be present; other carbon compounds and perhaps, elemental carbon are thought to be present.

For reasons that remain obscure—perhaps shock waves from a nearby exploding supernova*—star formation begins when a molecular cloud begins to collapse towards a local center of gravity. As the gas and dust of the great cloud fall towards the center of gravity, they heat up. Why? Because gravitational potential energy is converted to heat.

Energy! Among the towering achievements of nineteenth-century science, none is more magnificent or profound than the concept of energy and the laws of thermodynamics. According to the *First Law of Thermodynamics* there is something called energy in the universe that can be neither created nor destroyed. Defined as

*Supernovae are old stars that explode and, for a few months, shine with extraordinary brightness.

the capacity to do work,* energy can only be changed into different forms. A match, for example, contains a certain quantity of chemical potential energy; strike it, and you change that chemical potential energy into heat and light. Similarly, a brick held overhead has a certain amount of gravitational potential energy; drop it, and that energy is converted into sound, mechanical energy, and heat. A great deal of heat would be generated by dropping something as large as the Sun!

The collapsing cloud becomes a star when the material in the center grows so dense and hot that simple hydrogen nuclei (protons) fuse together, in a series of steps, to form helium nuclei. The radiant energy released from this thermonuclear fusion provides the back pressure that balances the infalling mass of the star. As the young star glows within, the gathering cloud of gas and dust around it settles into a disk-shaped region centered on the star's equatorial plane. Eventually the dust and gas begin to cool.

As the nebula cools, chemical compounds condense and coalesce, just as snow flakes form in Earth's atmosphere. Because the nebula is hotter and less oxidizing near it and colder and more oxidizing far from the star, the nature of the condensates varies from one concentric region to another. Around our Sun, in the region where the terrestrial planets formed, it seems that compounds of iron, silicon, oxygen, magnesium, and calcium were prominent among the early condensates.

The primitive Earth was formed by gravitational accretion of the condensed chemical compounds. These accreted compounds were enriched, relative to the sun, in elements heavier than carbon. Among the earliest, highest-temperature condensates we would expect oxides (e.g., aluminum oxide (Al_2O_3)—the mineral corundum; and calcium titanium oxide ($CaTiO_3$)—the mineral perovskite), silicates (e.g., calcium magnesium silicate ($CaMgSi_2O_6$)—the mineral diopside; and magnesium silicate (Mg_2SiO_4)—the mineral olivine), and iron metal. As the temperature dropped, these condensates

*Thermodynamics, which is the study of energy, its forms and transformations, defines work precisely and in a more restricted sense than the colloquial meaning: "When a force acts against resistance to produce motion in a body, the force is said to do work. Work is measured by the product of the force acting and the distance moved through against the resistance."[10]

would be joined by others (e.g., iron sulfide (FeS)—the mineral troilite; and magnetic iron oxide (Fe_3O_4)—the mineral magnetite).

As a physical process, *accretion* is hard to imagine, especially in the early stages, because most of us are not familiar with phenomena occurring in the absence of a powerful gravitational force. Dust grains usually settle on Earth before they stick together. Snowflakes and hailstones, however, accrete (stick together) in the atmosphere before falling to the ground. The accretionary process begins with minute grains of dust sticking together, then larger bits, then larger, and eventually a few really big pieces banging together. Some of these tremendous late collisions would have been disruptive, breaking objects apart rather than building larger masses. The Moon, for example, may have been formed by a collision between Earth and a Mars-sized object. The different spins observed among the planets today may have resulted from different impact histories as the planets grew. Unusual impacts might explain why Venus, Uranus, and Pluto spin backwards, why Venus turns so slowly, and why Uranus rolls along on a nearly horizontal spin axis.

Until the Hubble Space Telescope was launched (and repaired) planet-forming processes had never been observed. With the Hubble, however, astronomers have seen clouds of dust swirling around young stars, just as protoplanetary dust must once have surrounded the Sun. Such circumstellar disks of dust are, in fact, commonplace. A survey of young stars in the Orion nebula suggests that at least half of them are immersed in dust clouds. Before long, direct observations of circumstellar clouds will constrain and inform our theories of planet formation.

* * *

Although we have a good idea of Earth's bulk chemical composition, we do not know whether or not the original planet was chemically homogeneous. We presume it to have been solid, with an average temperature of 1,000°C. Why was Earth so hot? Mostly because of the conversion of gravitational potential energy to heat. Much of the Earth's early heat came from the energy of infalling objects, large and small.

If the original planet was chemically homogeneous, within a few hundred million years radioactive decay of short-lived atomic nuclei would have raised the average temperature of the interior from about 1,000°C to about 2,000°C and might have brought about the event known as the *Iron Catastrophe*. This hypothetical event, first proposed by F. A. Vening Meinesz in the early 1950s and subsequently considered by Harold Urey and Walter Elsasser, would have melted and rearranged Earth into chemically distinct zones, including a core of mostly molten iron. The catastrophe is presumed to have started when melting created droplets of molten iron, the density of which is roughly twice the average density of the planet. Thus, if droplets of iron appeared in the Earth, they would have immediately sunk towards the center. By sinking they would have converted gravitational potential energy to heat in a self-perpetuating process until all of the iron metal had settled to the center of the planet. The Earth's core contains iron plus some other less abundant elements (probably including hydrogen, carbon, oxygen, sulfur, and nickel) that, according to the model, were either swept along or actually dissolved in the molten iron. So much heat would have been generated by the catastrophe that much, if not most, of the planet would have melted. As it cooled, Earth would have been reorganized (differentiated) into the major spherical shells we find today: an iron core; a mantle of dense silicates and oxides; and a crust of less dense silicates, oxides, and other compounds.

Is this idea of pervasive planetary melting and differentiation correct? Perhaps not. Some chemical data are quite inconsistent with the simple model described above. In particular, the composition of the outer mantle zone seems too rich in nickel to have ever been in contact with molten iron, which would have dissolved the nickel and carried it into the core. Numerous attempts by experimentalists and theoreticians to find ways out of this bind, by resorting to high pressures for example, have proved fruitless. At this time the idea seems to have failed an important test and must be replaced by a different idea that does not derive core material from the outer mantle.

Regardless of whether the Iron Catastrophe is valid or must be replaced by a different idea, the differentiation of the planet into three general zones is true. These zones are the crust, the mantle, and the core, and they have the following dimensions:

Unit	*Thickness*	*Mass*
Crust	15–50 km	2.5×10^{25} g
Mantle	2,850–2,900 km	4.05×10^{27} g
Core	3,600 km	1.9×1027 g

This chemical differentiation concentrated lighter elements with lower melting temperatures, combined with silicon and oxygen, in the crust; heavier, higher-melting elements, again combined with silicon and oxygen, in the mantle; and iron in the core. Earth contains sufficient internal energy to circulate these materials. Consequently, Earth is dynamic, its physical transformation and chemical differentiation, at the surface and below, unceasing even today. Some of the energy for this process is from the planet's original heat, and some is from ongoing conversion of gravitational energy to heat (heavy elements sinking and light elements rising), but most is from decay of longer-lived radioactive atoms.

Earth has another set of concentric zones, superimposed upon the chemical zones, that are the result of temperature and pressure changes and the consequent differences in physical behavior. The outer zone, comprising 100 kilometers of the planet, is called the *lithosphere* because it is cool enough to behave as a strong, rigid shell. The lithosphere contains all of the crust and part of the outermost mantle. Beneath the lithosphere and extending down several hundred kilometers is the *asthenosphere*. This zone is weaker and less rigid than the lithosphere above or the lower mantle below. Many of the materials in it are close to their melting temperatures, and some are, in fact, molten. The base of the asthenosphere is not as clearly defined as its top, but it occurs generally about 350 kilometers below the surface. From the base of the asthenosphere down to the core, at a depth of 2,900 kilometers, the lower mantle is composed of similar chemical elements as the overlying zones (i.e., metals such as iron, magnesium, and calcium combined with oxygen and silicon), but they are packed together in extraordinarily dense arrangements that are stable only under exceedingly high pressures.

* * *

Earth's most remarkable feature, of course, is life. The appearance and evolution of life on Earth has been favored by three terrestrial characteristics: first, the planet evolved in such a way as to accumulate vast quantities of H_2O on the surface; second, Earth's surface temperature and pressure are close to the point where water, ice, and water vapor coexist; and, third, harmful radiation and high-energy particles from the Sun and elsewhere are mostly absorbed or deflected by the upper atmosphere.

Earth's ocean and atmosphere have evolved throughout the history of the planet, and continue to change today. Their original source and composition are not yet clearly identified. Compositional characteristics, such as less than solar proportions of the inert gases neon, krypton, and xenon, indicate that our atmosphere did not develop directly from nebular gas, although there may be a small contribution from that source. Two other sources seem to have been dominant: first, the accreting solids , ranging from dust grains to moon-sized objects (*planetesimals*), which would have carried hydrous compounds and other volatile-bearing substances into the growing planet; later, impacting meteors and comets, which would have delivered substantial additional supplies of volatiles, especially water, carbon, and nitrogen, to the Earth's surface. We do not yet know the relative importance of these two sources. The traditional view, greatly strengthened by the work of American geologist William Rubey in the 1950s, holds that the water and other volatiles were accreted with the planet and have been brought to the surface over geologic time by volcanic eruptions. Over the past 30 years,

Carboniferous vegetation.[11]

however, we have come to appreciate the potential of impacting comets as a continuing source of water and other volatile substances for Earth. We have also learned that volatile substances, such as water and carbon dioxide, are cycled repeatedly through Earth's interior by the planet-wide circulation process known as plate tectonics.

Earth's ocean and atmosphere are absolutely crucial to life, but we should bear in mind that in spite of their broad distribution on Earth's surface, the ocean and atmosphere comprise less then 0.025 percent of the mass of the planet. They are, as one of my favorite professors once said, "a very small tail on a very large dog." The other 99.98 percent of Earth is rock and molten iron.

Two

Atoms, Crystals, and Rocks

The study of rocks and minerals is one of the taproots of science, perhaps not as deep as astronomy, but extending well back into prehistory. Paleolithic people recognized and sought specific rocks and minerals for cutting tools, pigments, and various practical, religious, or artistic objects. The first minerals that were known and sought for their favorable properties were microcrystalline* varieties of quartz, particularly flint. The first-known miners dug for flint nodules buried in European chalk beds. Deep within caves at Lascaux, Font-de-Gaume, and Altamira, Paleolithic artists drew brilliant dream wishes of fat bison, deer, and other game with a vivid, multihued palette of pigments based upon mineral powders. Red, yellow, and brown iron oxides and black manganese oxide, the minerals hematite, goethite, limonite, and pyrolusite, respectively, were staples, but the list of substances included a dozen or so other minerals as well.

Native metals—copper, gold, silver, and iron—were sought early on for their beauty and usefulness. The earliest was copper in southern Turkey about 7,000 B.C. Colorful copper-bearing minerals such as green malachite, blue azurite, and brassy chalcopyrite joined the list of valued substances as the technology of copper progressed from shaping natural pieces to melting and casting the

*Microcrystalline means too small to be seen without a microscope.

metal and then to smelting the metal from minerals that were often associated with it. Eventually, around 2,500 B.C., the smelters discovered the benefits of adding minerals bearing arsenic, antimony, or tin to copper and thus ushered in the Bronze Age.

Native iron, that is, natural iron metal, is far more rare on the face of the Earth than copper, silver or gold, and so it was originally regarded as a precious substance. Most ancient iron artifacts were derived from meteorites, although the oldest known artifact, from a 5,000 B.C. grave in northern Iraq, was smelted (perhaps accidently) from iron ore. Purposeful iron smelting burst upon the world in a technological revolution around 1,000 B.C., and since then iron has been the single most important material in human civilization, its production growing exponentially through the years to more than a billion metric tons per year in the 1990s.

Mineralogy and its cousin, chemistry, progressed millenium after millenium, gathering experience and improving recipes for useful materials, but seldom advancing from empirical knowledge to potent theory. The difference is important. Empirical knowledge, based upon experience, is useful for predicting what will happen in the known world. Theory has the potential of leading us further, into worlds as yet unknown. The study of rocks and minerals did not begin to develop a theoretical basis until the seventeenth century because, unlike astronomy, the study of rocks and minerals was seldom quantitative.

In 1669 Nicolaus Steno discovered that similar faces of quartz crystals always bear the same angular relation to each other. Christened Niels Steensen, the son of a Danish goldsmith, Steno had studied medicine, latinized his name, and made his living as a physician. In 1665 he moved to the Court of Grand Duke Ferdinand II in Florence, which was a center of creativity and intellectual activity. There, in addition to his medical studies, Steno extended his attention to crystals and rocks.

To understand what Steno discovered, imagine a shoe box, which has three sets of similar surfaces: top/bottom, ends, and sides. The angular relationships between the faces are defined as follows:

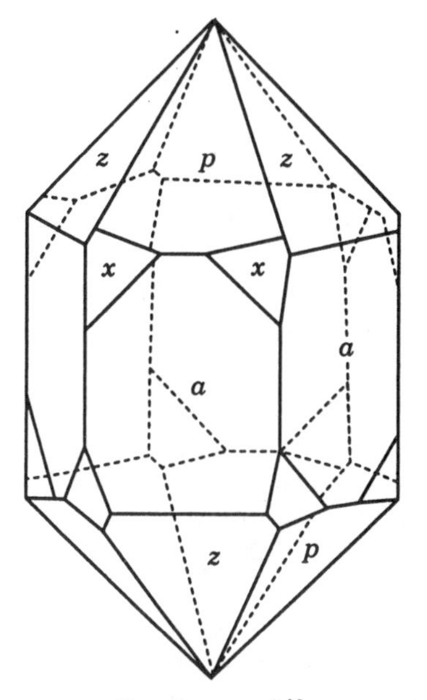

Quartz crystal.[12]

The angle between the top or bottom and a side is 90 degrees; the angle between an end and a side is 90 degrees; and so forth. Now transform the shoe box into a railroad boxcar and identify the sides of the boxcar that correspond to the top/bottom, ends, and sides of the shoebox. Do you see that the same angular relations obtain between similar faces of shoeboxes and boxcars, in spite of the fact that the objects are very different in size and not exactly the same shape? That is what Steno discovered for quartz crystals, which are not shaped like shoe boxes but, rather, like hexagonal prisms with pointed ends.

Steno's observation preceded any other evidence of internal regularity in crystals and led eventually to a general principle known as *Steno's Law,* which states that the angles between equivalent faces of crystals of the same substance, measured at the same temperature, are constant. Steno did not realize that what was true of quartz was true of all crystalline compounds. That fact was discovered in the next century by a French laboratory assistant named Arnould Carangeot (1742–1806). Although Carangeot has been largely forgotten, the general principle that he discovered and the measuring device that he invented* were important contributions to the study of crystalline matter.

The story of Steno's Law and Carangeot illustrates a fundamental difference between science and art. Art is created by individuals and is a unique, individual achievement. Science, in contrast, is a communal venture wherein ideas evolve and unprecedented ideas are extremely rare. Successful ideas acquire the names of leading individuals as labels, and the honor of association is generally well deserved. However, when their histories are examined almost all ideas are found to be the work of many minds over a period of time, and although it would be misleading and disrespectful to ignore the role of extraordinary individuals in science, their contributions (unlike the irreplaceable contributions of a Picasso or a Mozart) could *and would* have been made by others. C. P. Snow, writing of Ernest Rutherford, referred to "the anonymous Tolstoyan nature of science—how many years' difference would it have made if he had never lived? How much longer before the nucleus would have been understood as we now understand it? Perhaps ten years. More likely only five."[13]

> A contemporary poet has characterized this sense of the personality of art and of the impersonality of science in these words—"Art is myself; science is ourselves."
>
> *Claude Bernard*[14]

What sets the leaders in science apart from the rest? Of course, intellectual brilliance and access to learning are required, but beyond these prerequisites, most leaders are characterized by passion, productivity, and an apparently instinctive sense for choosing targets. An excellent example is the man known as the "Father of Modern Mineralogy," René Just Haüy.

Abbé René Just Haüy (1743–1822) was a cleric from a family of modest means who became passionately interested in mineralogy. In

*Carangeot invented the contact goniometer.

1784 he published a book in which he told of visiting a friend who was also a mineral collector. Haüy's friend handed him a cluster of calcite crystals ($CaCO_3$), which the Abbé dropped, causing it to smash to smithereens. After recovering from his embarrassment, Haüy observed that the broken (cleavage) fragments of calcite all had the same rhombohedral shape. He continued to smash them and found, to his amazed delight, that no matter how small he made them, the rhombs persisted.* He then went home and smashed his own collection of calcite crystals, with the same result. From this experience Haüy inferred that all crystals are made of small polyhedral units, the unit for each mineral having a characteristic shape. This was not an unprecedented observation or idea, but when Haüy went on to define the mathematical relationships between crystal forms and the fundamental polyhedral units, he founded modern crystallography.

Beginning with the fundamental polyhedral unit for a particular crystalline substance, Haüy defined reference axes by which to describe the positions of the faces of a crystal of that substance. He then found that the crystal faces always cut the reference axes at simple rational multiples of specific lengths. This observation, which is known as the *Law of Simple Rational Intercepts*, or the *Law of Haüy*, demonstrates that crystals are constructed by repeating simple three-dimensional units. We now understand that these units are not atoms but three-dimensional arrangements of small numbers of atoms.

The concept of atoms as indivisable, fundamental units of matter had come from ancient Greece, but their reality was not generally accepted by philosophers or scientists until the beginning of the twentieth century. René Descartes (1596–1650) was prominent among the nonbelievers:

> We also know that there cannot be any atoms or parts of matter which are indivisable of their own nature (as certain philosophers have imagined). For however small the parts are supposed to be, yet because they are necessarily

*A rhombohedron, or rhomb, is a regular six-sided prism, like a shoe box, but, unlike a shoe box, its six sides are parallelograms rather than rectangles.

> extended we are always able in thought to divide any one of them into two or more parts; and thus we know that they are divisable.[15]

In spite of their problematical status as real objects, however, atoms became important theoretical entities in 1805, when John Dalton (1766–1844) enunciated his *Chemical Atomic Theory.*

Dalton's theory, which is the cornerstone of quantitative chemistry, came in a period of extraordinary intellectual vigor in western Europe, Britain, and America, in which political, industrial and intellectual revolutions burst like skyrockets over the landscape of human history. For chemistry, especially, it was a time of brilliance, exhibited by such men as Avogadro (1776–1856), Berzelius (1779–1848), Cavendish (1731–1810), Davy (1778–1819), Faraday (1791–1867), Gay-Lussac (1778–1850), Lavoisier (1743–1794), Priestley (1733–1804), Scheele (1742–1786), and Volta (1745 –1827). Sciences tend to advance in rushes and then consolidate their knowledge for a somewhat longer period before the next rush. While the rush is on, the science attracts the brightest and most energetic minds. Such was the case of chemistry at the turn of the nineteenth century.

In the city of Manchester, England, at the heart of the industrial revolution, John Dalton deduced (from weights of chemical reactants and reaction products) that elements are made of minute, discrete, indivisible particles that maintain their identity through all chemical and physical changes; that atoms of the same element are identical; that chemical combinations between two or more elements involve simple numerical ratios of atoms to form molecules; and that atoms of an element can unite in more than one ratio to form a variety of compounds. These atoms were controversial and something of an embarrassment. Yes, they led to clearer understandings of chemical reactions, but many scientists doubted their existence. How could imaginary objects elucidate Nature? Similar situations developed with antimatter in the 1930s and quarks in the 1960s. As I said, theories often lead to worlds as yet unknown.

For bright-eyed creatures such as ourselves, light is not merely a metaphor of knowledge, it is a principal means of knowing. Thus, like so much else, our understanding of crystals is intimately related to our understanding of light. Newton viewed light rays as particle (corpuscle) beams. His corpuscular theory was challenged by Dutch astronomer and physicist Christian Huygens (1629–1695), whose *Traité de la Lumiére* in 1669 presented light as a wave phenomemon.

During the early nineteenth century the duality of light, as particulate and wave phenomena, attracted the attention of numerous physicists. Foremost among these was Thomas Young (1773–1829), a person of legendary precocity, intelligence, and accomplishment, who was admitted to the Royal Society at age twenty-one. Among his achievements were his fluency in numerous languages as a child, his discovery of the focusing and color sensing mechanisms of the human eye, and his crucial contributions to the deciphering of the Rosetta Stone. He also discovered the phenomenon of the interference of light, which he summarized as follows:

> . . . wherever two portions of the same light arrive at the eye by different routes, either exactly or very nearly in the same direction, the light becomes most intense when the difference in the routes is any multiple of a certain length, and least intense in the intermediate state of the interfering portions; and this length is different for light of different colors.[16]

The length Young referred to is the wavelength. The conditions of maximum and minimum intensity (brightness) are caused by parallel light waves that combine either to increase their combined brightness (constructive interference) or to decrease their combined brightness (destructive interference). Whether interference between waves is constructive or destructive depends upon the timing of their crests and troughs. If both waves crest at the same time their interference is constructive, but if one wave crests while the other is in a trough, their interference is destructive.

If waves A and B are combined, the intensity (amplitude) increases to a maximum:

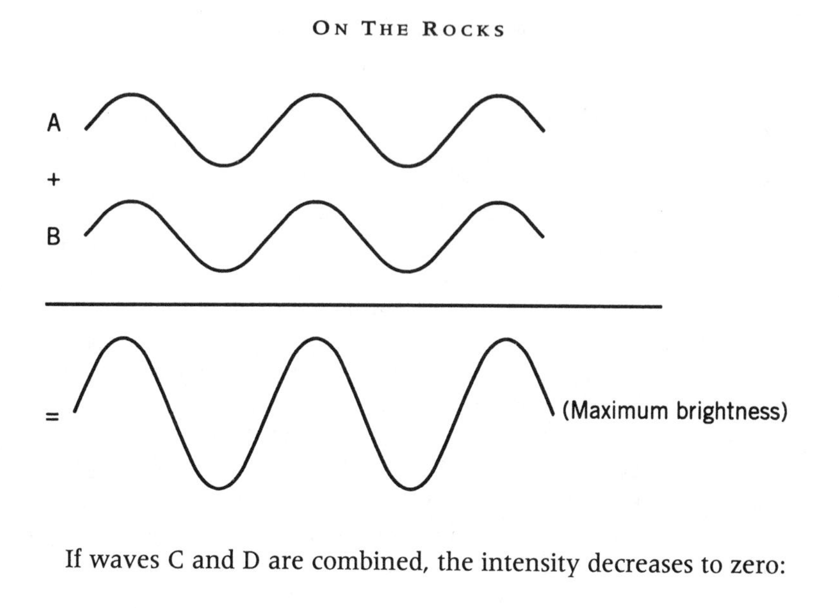

If waves C and D are combined, the intensity decreases to zero:

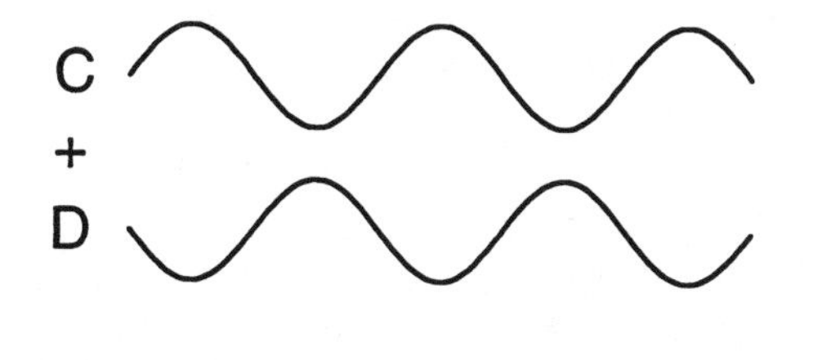

Experiments by Young and his contemporaries involving interference phenomena provided physical evidence of the internal regularity of crystals on the scale of the wavelengths of visible light (10^{-5} cm). A century later Young's explanation of interference would be used to demonstrate the existence and regular arrangement of atoms in crystals.

In 1895 the German physicist Wilhelm Röntgen (1845–1923) discovered X rays. The discovery occurred "by accident," but Röntgen was prepared to take advantage of the accident, in a good example of serendipity.* Unlike most scientific discoveries, this one really did have the setting and drama of a grade-B science fiction movie.

Herr Professor Röntgen was a man with intense dark eyes and a formidable black beard who liked to work long, late, and alone in his laboratory. His researches had been wide-ranging and substantial, and he was a skilled and indefatigable experimentalist with 48 published papers. In June of 1894 he began experimenting with electrical discharges through partially evacuated glass tubes. (Such tubes are called *cathode ray tubes* because of the rays that are emitted from the negatively charged electrode, or cathode.) By 1895 he had become interested in weak fluorescence, which seemed to be caused by cathode rays in the immediate vicinity of charged tubes that had been enclosed in opaque cardboard shields. Late one November afternoon he wrapped a tube in cardboard so that no light could go in or out, pulled the drapes so that the laboratory was pitch dark, and turned on the electricity. Although the tube was charged, and surely glowing inside the cardboard, no light leaked through. Suddenly, out of the corner of his eye, across the darkened room, Röntgen noticed a faint green light. Imagine his surprise! When he turned the tube off, the light disappeared. When he turned the tube back on, the light reappeared. He struck a match to see. The glowing object was a fluorescent screen lying on a table far beyond the range of a normal cathode ray.

Over the next few days Röntgen experimented with this mysterious phenomenon, finding that it behaved like radiation but had unprecedented penetrative power. It could pass almost undiminished through stacks of cloth or paper, but some materials could stop it. Then he held a lead disk in front of a fluorescing screen and saw its shadow—and something more appalling. He saw the faint shadows of his thumb and forefinger holding the lead disk and, more distinctly, the shadows of his living bones!

The remaining weeks of 1895 were a time of extreme intensity and effort; by the end of December he had prepared an initial report, "Ueber eine neue Art von Strahlen" ("Concerning A New

*Once upon a time, according to a Persian fairy tale, three princes from Sri Lanka, or Serendip as the Persians called it, displayed a remarkable talent for finding treasures of one sort or another by surprise.

Kind of Rays"), whose first paragraph is a model of clarity and precision. Anyone familiar with the apparatus could read his description and reproduce his experiment:

> If the discharge of a fairly large Ruhmkorff induction coil is allowed to pass through a Hirttorf vacuum tube, or a sufficiently evacuated Lenard, Crookes tube, or similar apparatus, and if one covers the tube with a fairly close fitting mantle of thin black cardboard, one observes in a completely darkened room that a paper screen painted with barium platinocyanide placed near the apparatus glows brightly or becomes fluorescent with each discharge, regardless of whether the coated surface or the other side is turned towards the discharge tube. This fluorescence is still visible at a distance of two meters from the apparatus.[17]

Although he won the first Nobel prize for the discovery of X rays, Röntgen did not know what they were. That discovery, and another Nobel prize, went to another German physicist, Max von Laue (1879–1960), who designed an experiment in 1912 to test the proposition that X rays were high-energy electromagnetic waves of very short wavelength. Von Laue assumed that the atoms in crystals are regularly spaced and that the spacing is of the same order of magnitude as the wavelength of X rays. If that were so, the crystal would diffract the X-ray beam.

Everyone knows what reflection is, and most people know what refraction is. However, few people know what diffraction is. *Diffraction* is "any deviation of light rays from rectilinear paths which cannot be interpreted as reflection or refraction."[18] One of the most obvious displays of this phenomenon occurs when sunlight shines through the parallel chain link fences enclosing a pedestrian overpass on a highway. The image is distorted by a series of diagonal, wavy, bright and dark bands, which are caused by diffraction of the light passing through gaps between the superimposed wire grids. Constructive and destructive interference occurs because the gaps are of the same order of magnitude as the wavelength of the light. Another way to observe diffraction is to observe the bright sky through the gap between two razor blades or credit cards. Just before the blades or cards touch, the image of the edges and the intervening gap breaks up into several dark and light lines. That, too, is diffraction.

Von Laue told his assistants, Walter Friedrich and Paul Knipping, to photograph an X-ray beam through a crystal of copper sulfate. The result was a pattern of dots resulting from diffraction of the X rays, which confirmed both the short wavelength of X rays and the orderly arrangement and close spacing of the atoms. This experiment opened the field of X-ray crystallography by which the internal structures of hundreds of thousands of crystalline solids have been determined.

A mineral is a naturally occurring, solid chemical compound characterized by a definite composition (or range of composition) and a specific atomic structure. This definition is a useful convention, but it does create some awkward situations. Volcanic glass is not a mineral because its composition is variable and its structure is disordered. Ice is a mineral, but water is not, nor is native mercury. In spite of these awkward situations the study of minerals is the first step in the study of rocks.

The different structures of crystals and their compositional limits can be understood by considering the differences between atoms of different elements. Since the invention of the scanning tunneling microscope (STM) in 1981 by Gerd Karl Binnig and Heinrich Rohrer, it has been possible to visualize and manipulate(!) individual atoms. Atoms had been imagined as roundish, fuzzy objects of varying sizes, composed of a positively charged nucleus surrounded by a cloud of negatively charged electrons. STM images confirmed this traditional view and revealed the beauty of Nature, on the atomic scale, in breathtaking style.

The invention of scanning tunneling microscopy, for which Binnig and Rohrer received a Nobel Prize, was a technical advance of supreme importance. It was also contrary to conventional wisdom. Of the experience Rohrer has written,

> STM was not developed from one of the already existing local probe methods or from ideas about them, nor was it

> done in the community of microscopists or in other circles with the appropriate competence. No technically new component or new material was necessary, no new physical insight was required and no additional theoretical basis had to be established, yet somehow the belief prevailed in these communities that "it" could not be done. . . . We heard so many objections, for example to the positioning of a local probe with subångstrom accuracy, including objections citing the uncertainty principle, even after the STM had worked! We might learn from this that an occasional change in the field of interest can bring unexpected progress."[19]

The size and behavior of an atom depends upon the nature of the electron cloud associated with it. Chemical bonds occur when adjacent atoms share electrons, or when positively and negatively charged atoms, called *ions*, attract each other. Atoms share electrons, or acquire electrostatic charges, because certain discrete numbers of electrons in the outermost, or *valence*, shell of the electron cloud make the atom or ion more stable, that is, less likely to engage in a chemical reaction.

Uncharged atoms have one to eight electrons in their outer shells. In many chemical compounds, including minerals, the various atoms, except for hydrogen, achieve stability (or saturation) when they acquire eight valence electrons. This simple "octet rule" has many interesting exceptions, but it is sufficient for our purposes. Alkali metals (e.g., sodium or potassium), with one valence electron, tend to lose that electron and thereby achieve a stable outer shell. Having lost one electron, the atom has a positive charge and is called a *cation*. The halogens (e.g., fluorine or chlorine), with seven valence electrons, tend to acquire one electron in order to have eight in the outer shell. Having added one electron, the atom has a negative charge and is called an *anion*. Elements between the alkali metals and the halogens may lose or acquire two or more electrons in order to achieve a stable octet, but, as their requirements increase, atoms tend to share electrons rather than lose or acquire them completely.

Ionic bonds involve oppositely charged ions; the ions are discrete and exist as ions in solution or when the compound is molten. *Cova-*

lent bonds occur when the electron clouds of adjacent atoms overlap and some electrons are shared between the two atoms; some of these bonds may persist when the compound is in solution or molten. Two common substances furnish examples of each type of bonding. Sodium ions (Na+) and chloride ions (Cl^-) form common table salt (sodium chloride), with ionic bonds. As an example of covalent bonding, consider water wherein one oxygen atom, with six valence electrons, shares electrons with two hydrogen atoms, each of which has one valence electron. By sharing, the hydrogen atoms each attain stability with two valence electrons, and the oxygen atom attains its stable configuration with eight valence electrons.

Because there are only eight possible arrangements of electrons in the valence shell, chemical and physical properties are repeated regularly among the elements. This repetition, or *periodicity*, of properties was discovered in 1869 by Dmitri Ivanovich Mendeleev (1934–1907). Long before there was any understanding of atomic structure, nearly thirty years before J. J. Thompson (1856–1940) discovered the electron, Mendeleev tabulated the known chemical elements according to their atomic weights and found that properties recurred in cycles of seven. (The inert gases of Group Eight had not yet been discovered.)

The first 3 periods include the first 18 elements, hydrogen through argon. The number of valence electrons places each of these elements in a column, or group, of similar elements. Although the table becomes somewhat more complicated in subsequent periods, the fundamental arrangement of elements into successive periods of eight chemically similar groups remains.

	Number of Outer (Valence) Electrons							
Period ↓	*1*	*2*	*3*	*4*	*5*	*6*	*7*	*8*
1	H							He*
2	Li	Be	B	C	N	O	F	Ne
3	Na	Mg	Al	Si	P	S	Cl	Ar

*Helium has only two electrons.

By noting an element's position in Mendeleev's periodic table, we can predict whether it will tend to form ionic or covalent bonds and which elements are likely to have similar properties. Atoms of elements on the far right tend to pick up electrons, and atoms of elements on the left tend to lose them; atoms in between tend to share. The inert gas atoms of Group Eight on the far right already have eight valence electrons and are therefore unreactive.

The atomic proportions of chemical compounds, and their chemical properties, are determined by the electronic configurations of the constituent atoms. How atoms or ions fit together in crystals is determined by their electronic configurations and by their relative sizes. Size considerations in dominantly ionic compounds can be understood by regarding the ions as hard spheres. How many spheres of one size can you squeeze around spheres of another size? This is called the *coordination number*, and it depends upon the relative sizes of the spheres.

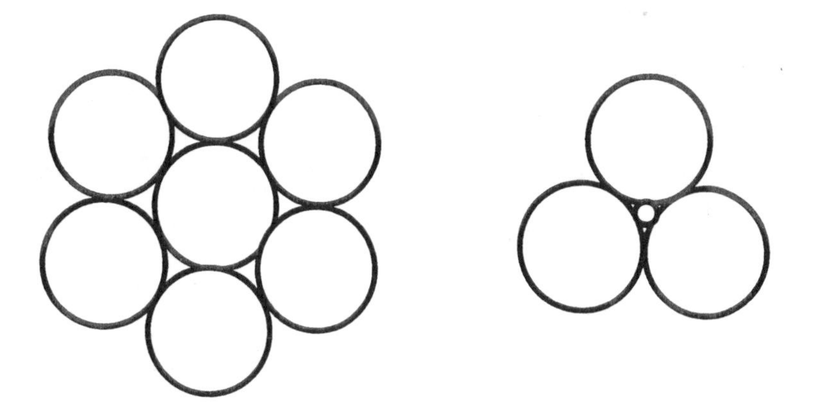

A quick way to appreciate the coordination number is to play with coins. Coins of the same size can have six nearest neighbors—six quarters can fit tightly around one quarter. If you substitute a smaller coin in the center, the coordination number goes down—

only five quarters can fit around a dime. Notice how much smaller a coin would have to be to fit tightly between three quarters. A quarter is about an inch in diameter, so the smaller coin would be only 0.14 inch in diameter. We express relative size differences as the *radius ratio,* which is the smaller radius divided by the larger radius. Thus, the radius ratio of our tiny coin between three quarters would be 0.14. The geometry becomes a bit more complicated when the objects are spheres instead of circles, but the principal is exactly the same. When the spheres are really ions, the radius ratio is usually the cation radius (the dime) divided by the anion radius (the quarter).

Radius Ratio	*Coordination Number*	*Arrangement of Atoms*
0.14 to 0.22	3	3 around 1: triangular
0.22 to 0.41	4	4 around 1: tetrahedral
0.41 to 0.73	6	6 around 1: octahedral
0.73 to 1	8	8 around 1: cubic

Oxygen and silicon are the two most abundant elements in Earth's crust and mantle. Thus, the two most abundant chemical compounds in the crust and mantle are oxides and silicates. Six other elements characterize the most common varieties of these oxides and silicates. They are aluminum, iron, magnesium, calcium, sodium, and potassium. In Earth's crust, which is the only part of the planet we usually get to touch, these elements are present in the following proportions.

Composition of the Crust[20]

Element	*Weight (%)*	*Atomic (%)*	*Volume (%)*
Oxygen	46.60	62.55	93.77
Silicon	27.72	21.22	0.86
Aluminum	8.13	6.47	0.47
Iron	5.00	1.92	0.43
Magnesium	2.09	1.84	0.29
Calcium	3.63	1.94	1.03
Sodium	2.83	2.64	1.32
Potassium	2.59	1.42	1.83

Oxides are made of oxygen atoms packed around smaller metal atoms. If the metal atoms are small enough to fit between four or six oxygen atoms, the oxygen atoms can fit together in the tightest possible packing arrangement. Known as *close packing,* this arrangement has each oxygen atom surrounded by six others in the same layer, seated in a cusp formed by three oxygen atoms in the layer below and capped by three oxygens in the layer above. Within this pile of spheres, which resembles a pile of cannon balls in the park, there are tiny spaces surrounded by either four or six oxygen atoms which are occupied by metal atoms. How many oxygens surround each metal atom depends upon the radius ratio. When the metal is aluminum, the radius ratio is large enough for each aluminum atom to be surrounded by six oxygens. This is the basic structural unit of the mineral corundum (Al_2O_3) as well as the iron oxide mineral hematite (Fe_2O_3). There are many other oxide minerals with metal atoms stuffed into the interstices of close-packed oxygen atoms.

Atoms of silicon, the second most abundant element in the crust and mantle, are smaller than those of aluminum or iron. The radius ratio with oxygen is small enough that only four oxygen atoms can surround one silicon atom, with the oxygen atoms located at the four corners of a tetrahedron. The SiO_4 tetrahedron is the fundamental unit of silicate minerals and silicate melts. It is called the *SiO_4 monomer.*

Each oxygen atom has six valence electrons and therefore requires two more in order to attain a stable octet:

e

e O e
e

e e

Each silicon atom has four valence electrons and therefore requires four more in order to attain a stable octet.

e

e Si e

e

When four oxygens surround one silicon atom in the SiO_4 monomer, each oxygen shares one electron with the silicon, and the silicon atom shares one electron with each oxygen. This produces a stable octet around the silicon atom, but each oxygen atom is still one short of the desired octet.

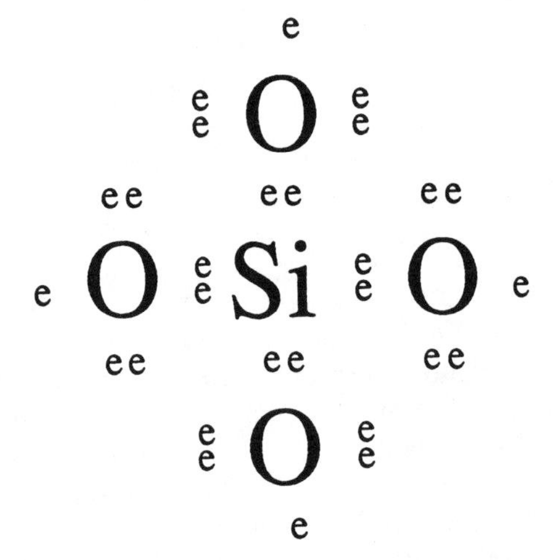

In silicate minerals, the four electrons may be obtained from other atoms in the structure, such as calcium, iron, or magnesium. These atoms then become positively charged ions or cations.

$$Mg = Mg^{2+} + 2e^-$$

The mineral olivine, $(Mg,Fe)_2SiO_4$, for example, consists of Mg^{2+} and/or Fe^{2+} ions and SiO_4^{4-} tetrahedra. Placing the symbols for magnesium and iron in parentheses, separated by a comma, indicates that olivine may contain either magnesium or iron or any mixture of the two elements. Another way to say this is that olivine is a *solid solution* of Mg_2SiO_4 and Fe_2SiO_4. Unlike other common silicates, the olivine structure is like an oxide structure—the oxygens are in a close-packed arrangement, with the iron and magnesium ions surrounded by six oxygens and the silicons surrounded by four.

Another way for an SiO_4 unit to acquire more electrons is by *polymerization,* a process where SiO_4 units share one or more of their oxygen atoms with other SiO_4 units:

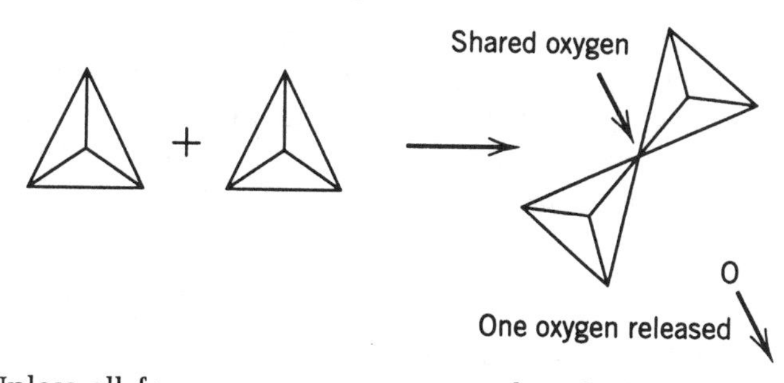

Unless all four corner oxygens are shared, the silicon-oxygen polymer still has a negative charge that must be balanced by positively charged metal ions such as sodium (Na^+), calcium (Ca^{2+}), or iron (Fe^{2+}).

In olivine the SiO_4^{4-} units are isolated—the silicon atoms are spread out so that they do not share oxygen atoms with other silicons. In other common silicates the SiO_4^{4-} units are polymerized. There are many different silicate polymers. The type determines whether the silicate will form *sheets,* like mica, or *blocks,* like feldspar. It also determines many of the physical properties, such as how hard the mineral is or how its crystals tend to break apart.

Pyroxenes are silicate solid solutions that may contain a variety of metals, such as calcium, iron, magnesium, sodium, aluminum, or chromium. They are constructed with chains of silicon-oxygen polymers, in which each SiO_4 unit shares two of its four corner oxygen atoms with other SiO_4 units.

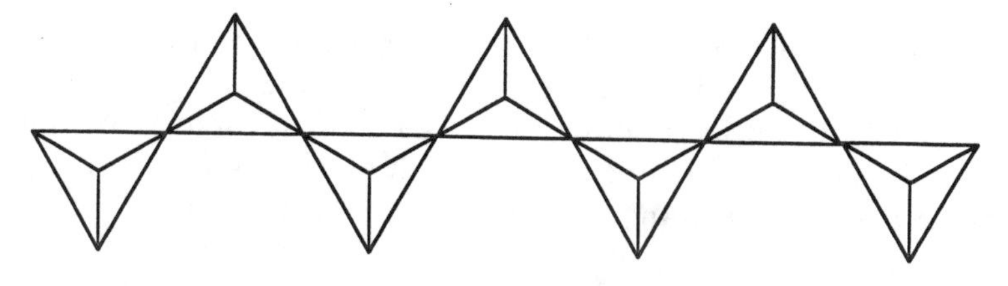

Some examples of simple pyroxene formulas are $(Mg,Fe)SiO_3$, $CaMgSi_2O_6$, and $NaAlSi_2O_6$. In natural pyroxenes these simple combinations, and others, may be dissolved together in one solid solution.

Even more complicated than pyroxenes are the minerals called *amphiboles*. These minerals, whose name means ambiguous, are built with bands or double chains of silicon-oxygen polymers. In addition to various metal atoms, amphiboles also contain hydroxyl (OH^-) and chloride or fluoride ions (Cl^- or F^-).

Note that the tetrahedra share two or three of their corners with other tetrahedra for an average of two and one half.

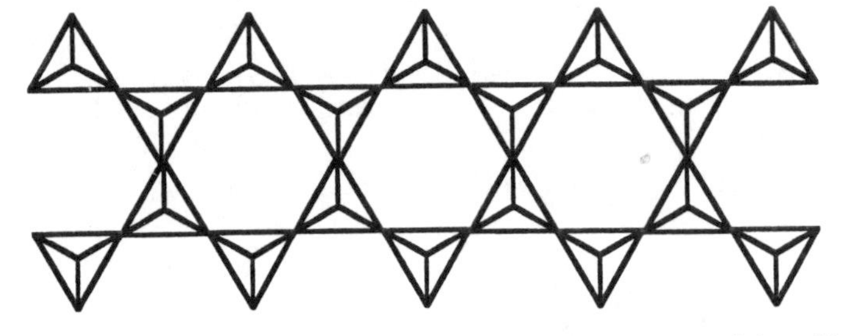

Micas, which you may have seen as window material in old-fashioned stoves or as insulating sheets in old-fashioned toasters, are based upon polymerized sheets of silicon and oxygen, with each SiO_4 tetrahedron sharing three of its four corner oxygens.

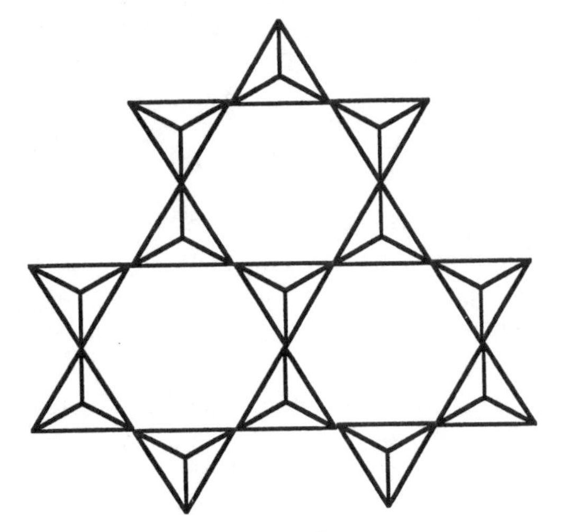

Quartz and feldspar, common minerals in Earth's crust, are based upon three-dimensional networks of silicon-oxygen polymers. In quartz, which is silicon dioxide (SiO_2), each SiO_4 unit shares all of its oxygen atoms with other SiO_4 units. The feldspars are complex solid solutions of sodium aluminum silicate ($NaAlSi_3O_8$) and potassium aluminum silicate ($KAlSi_3O_8$) *or* calcium aluminum silicate ($CaAl_2Si_2O_8$). The solid solutions of the sodium and potassium varieties are known as the *alkali feldspars,* and the solid solutions of the sodium and calcium varieties are known as the *plagioclase feldspars.*

A semiprecious silicate solid solution is January's birthstone, garnet. Like pyroxenes, garnets contain variable amounts of aluminum, chromium, iron, manganese, calcium, and magnesium. They form deeply colored (typically claret to burgundy), faceted spheres that are harder, stronger, and more dense than most other silicates. Garnets are found in many metamorphic rocks and are prominent constituents of Earth's mantle.

There are roughly thirty-five hundred minerals, among them many other compounds besides silicates and oxides. There are native elements such as gold. There are carbonates, sulfides, phosphates, and nitrates, and so forth: Halite is sodium chloride or common salt; pyrite, or fool's gold, is iron sulfide. Calcite and aragonite are two minerals with the same chemical composition, calcium carbonate, but with different structures. Calcite is the principal mineral in limestone, and aragonite is common in the shells of invertebrate organisms like clams and oysters. Magnetite is magnetic iron oxide (Fe_3O_4). This mineral was known to our ancestors as lodestone (meaning "lead stone") because it is strongly magnetic and can be used as a compass needle.

Graphite, one form of native carbon, is a minor constituent of many rocks in Earth's crust. Deeper down in the planet, carbon forms a very different mineral: diamond. Two minerals could not be more different. Graphite is soft and slippery; diamond is not slippery and is the hardest substance known. Graphite is opaque; diamond is transparent. Graphite conducts electricity; diamond is a very poor electrical conductor, although it is a better thermal conductor than graphite. These contrasting characteristics are due to the minerals' different crystal structures. Graphite has sheets of carbon atoms held together by bonds that are fairly strong in the plane of the sheet but very weak from sheet to sheet. Diamond has a strong three-dimensional structure, with each atom strongly bonded to four others that lie at the corners of a tetrahedron.

Chemists have synthesized far more compounds than there are minerals. Why aren't there more minerals? Because they must form under natural conditions and because most chemical compounds are not sufficiently stable to survive for long periods. Moreover, minerals tend to be compounds that are not likely to decompose spontaneously or to react with other natural substances around them. To this last statement ice and halite are striking exceptions.

A rock is a natural aggregate of mineral grains. It usually contains more than one kind of mineral, although some contain only one, and it is classified, according to its mode of formation, as sedimentary, igneous, or metamorphic. *Sedimentary* rocks are formed by accumulation of mineral grains in water or air; *igneous* rocks form from molten material called magma; and *metamorphic* rocks are igneous or sedimentary rocks that have been changed (metamorphosed) by changes in temperature, pressure, or chemical environment. Simple or complex, rocks are the documents of earth and planetary history.

If the greatest intellectual achievement of the earth sciences has been the concept and measurement of geologic time, the least praiseworthy may be rock classification—the chaos of names created, defined, modified, discarded, and resurrected by generations of hapless geologists. Of this situation one exasperated fellow cried:

> Can we seriously believe that petrology [the science of rocks] is the one branch of natural history whose practitioners are able to understand relations among sets of objects they are unable to classify, and often to name, unambiguously?[21]

The root problem is that, at our current level of knowledge, many rocks *are* ambiguous. Unlike rabbits and cows, many of them *can* interbreed: Both magmas and sediments can mix, and metamorphic materials are notorious mixtures. Also, the kinds of chemical compounds, or minerals, that make up rocks depend upon myriad variables, such as pressure, temperature, bulk chemistry, and rate of crystal growth. Finally, because rock materials tend to react slowly, chemical reactions within them seldom go to completion and therefore rocks commonly contain several generations of materials—the reactants as well as the products of successive chemical and physical changes. A rock is a complex document, potentially able to recall, at least in part, every change experienced since the moment it formed.

In the broadest of classifications there are *fragmental rocks,* which are bits and pieces of other rocks, and *precipitates,* which are assemblies of crystals precipitated from media such as seawater or molten rock. Within these great classes are sedimentary rocks, deposited from air or water on the Earth's surface; igneous rocks, formed from melts; and metamorphic rocks, which have been squeezed or sheared, cooked or cooled, or subjected to chemical changes since birth. Some rocks are rich in certain chemicals and some rich in others, and the same combinations of chemical elements may be organized into different minerals.

Do rock names describe rocks? Some do, but most don't. Some rock names are ancient. *Granite* may derive from the Latin *granum,* meaning grained. *Basalt* is even older and its origin more obscure: Pliny the Younger attributed the name to Egypt and Ethiopia; perhaps the source was the Ethiopian word *bsul,* meaning cooked. In any case, these words tell us little or nothing. Other rock names refer to places where the type examples of specific rocks were described, but, thanks to subsequent changes in definitions, the current definitions may no longer describe the rocks at the original locations! We will slip through this jungle of rock names without bothering to meet many of its inhabitants, but we must master a few names in order to understand Earth and other rocky objects in our solar system.

Lithology is the study of the physical character of rocks, including all of the compositional and textural variations they display. This study is, of course, exceedingly old, originating literally with Paleolithic people. Until the mid-nineteenth century, however, the study of rocks was limited to those large features that are apparent to the unaided eye. There is much to see in some rocks at that scale, or with the assistance of a magnifying glass; nevertheless, our knowledge and understanding of rocks took a giant step forward when we learned to look at rocks with microscopes. This branch of lithology is called *petrography*.

Petrography was born in Sheffield, England, in 1849, when Henry Clifton Sorby (1826–1908) ground a limestone chip to transparent thinness and examined it under a microscope equipped with polarizing prisms. The polarized light optics of individual crystals had been investigated by physicists interested in the nature of light, and some physicians had examined bone slices with microscopes. Until Sorby, however, no one had thought to examine rocks that way. Sorby's moment of truth must have been breathtaking: Dull, featureless stones were transformed into intricate assemblages of brilliantly colored crystals, kaleidoscopic images like stained glass windows but more complex and therefore more informative. He saw what no one before him had seen.

Not that anyone cared. Although the Sorbys had been in Sheffield since the time of Henry VIII they were not a prestigious family, and Sorby had no particularly impressive academic credentials. In fact, he rather avoided such trappings, saying: "I worked not to pass an examination but to qualify myself for a career of original investigation."[22]

Established academics made fun of Sorby's efforts. It was ridiculous, they said, to study mountains with microscopes, but, of course, Sorby was proved right in the end. Eventually any significant new technique for observing nature (a telescope, a microscope, or a super collider) produces an observational stampede to look at everything in the new way. Soon after Sorby described his experiments to the Geological Societies of London and France the stampede was on, and it lasted for more than half a century until virtually every kind of rock on Earth had been sliced, ground thin, examined, described, and classified.

This was the golden age of petrography, when scientists, particularly those in France, Britain, and Germany, first observed the extraordinary beauty and complexity of rocks revealed by the polarizing microscope. In science unprecedented observations are made every day—through application of the petrographic microscope, for example—but unprecedented ideas, such as the preparation of thin sections and their examination in polarized light, are exceedingly rare.

As scientists looked more closely at rocks, seeing the intricate microscopic relationships between different minerals, they realized that there were recurring relationships and antipathies. Some minerals commonly coexisted; others were never found together. Scientists saw textural features that looked the same, even when the materials were different. They saw fine-grained materials stuffed into the interstices between larger bits, and they saw materials arranged in concentric zones, like moats, around other kinds of material. A profound conclusion from these observations was that although many rocks are complicated, all are subject to general rules, most of which are laws governing physical and chemical equilibrium. *Equilibrium* can be thought of as a balancing act between opposing changes as time goes on. If saltwater and salt crystals are in equilibrium, for example, then we can assume that just as many dissolved sodium and chloride ions are combining to build the crystals as are dissolving out of them. Equilibrium is in the realm of thermodynamics.

I was dining some years ago with Walter Alvarez, a scientist friend and former classmate who, along with his Nobel-laureate father Luis, Frank Asaro, and Helen Michel, discovered the extraterrestrial impact that doomed the dinosaurs. Our conversation turned to religion, and I said that I found the First Law of Thermodynamics, and particularly the concept of energy, to be quite moving. Walter smiled and said that yes, the First Law was profound and mysterious, but for him the Second Law was even more inspiring because it involved Time. At that point I realized that I did not really understand the Second Law. Most people don't. This widespread ignorance drew scorn from C. P. Snow in his provocative and influential book, *The Two Cultures*:

> A good many times I have been present at gatherings of people who, by the standards of the traditional culture, are thought highly educated and who have with considerable gusto been expressing their incredulity at the illiteracy of

> scientists. Once or twice I have been provoked and have asked the company how many of them could describe the Second Law of Thermodynamics. The response was cold: it was also negative. Yet, I was asking something which is about the scientific equivalent of: *Have you read a work of Shakespeare's?*[23]

With all deference to the late Sir Charles, that was hyperbole. The Second Law is more subtle and more difficult for many people to grasp than most works of Shakespeare, and there are, I suspect, one or two scientists who could not spontaneously provide a lucid description of the Second Law that was also correct.

As a reminder, it helps to start with the amusing analogy invented by David Langford:[24]

> The First Law says, "You can't win."
> The Second Law says, "You can't break even."
> The Third Law says, "You can't get out of the game."

This flippant summary may not suffice if you encounter the likes of C. P. Snow at a cocktail party. A deeper understanding may be required.

The discipline called *thermodynamics,* which describes the distribution of energy under equilibrium conditions, was born of the Industrial Revolution and, particularly, the development of the steam engine, which began at the end of the seventeenth century. Engineers and inventors, striving to increase power production and efficiency, measured the performance of their machines with increasing precision using new instruments such as the mercury-in-glass thermometer that Fahrenheit invented in 1715. The laws of thermodynamics came from these measurements and related experiments.

Progress was delayed by the misconception that heat was a fluid, like water but without measurable mass, that flowed from a hot place (the boiler) to a cold place (the condenser) and drove the engine as water drives a water wheel. It took half a century of experiments and analysis to dispel this *Caloric Theory,* to replace it with the idea that heat is a form of atomic motion, and to summarize the relationships of heat and work in statements that are the laws of thermodynamics. The names of those who did this work reverberate in the history of science like the deep tolls of a great bell:

Benjamin Thompson (1753–1814), the Tory from Massachusetts who spied for the Crown, fled to Europe, and bored cannon for the Elector of Bavaria, who made him Graf (Count) von Rumford. While boring cannon Rumford realized that the heat associated with the process could not possibly be flowing from the metal—so much heat would have fused the billet—but must be related to the work of the drill. With this and other cogent observations Rumford sapped the Caloric Theory.

Sadi Carnot (1796–1832), the French military engineer who demonstrated that power output would be greatest if the machine operated in a cyclical fashion—always returning to its original condition—over the largest possible temperature range, and if each step could be performed reversibly—that, is so slowly that a minute change would reverse the step.

James Joule (1818–1889), the brewer's son from Manchester who learned his love of scientific truth and research from John Dalton. Joule and Julius Robert Mayer, a German ship's doctor, independently determined that work can be convertedly quantitatively into heat. By numerous careful experiments Joule measured the mechanical equivalent of heat, and the modern metric unit of energy or work is now called the *joule*.

Clausius, Clapeyron, Kelvin,* Boltzman, and Clerk Maxwell, the Titans of nineteenth century physics who wrote the thermodynamic laws.

The First Law defines energy and says that in any isolated system, such as the universe, energy may neither be created nor destroyed; it must be conserved. In the words of Rudolf Clausius, "Die Energie der Welt ist konstant."[25]

The Second Law declares that although work may be dissipated completely into heat, the reverse is not possible: Heat may not be converted entirely into work. The price to be paid when heat is converted into work is the change in *entropy*, defined by Clausius as the heat supplied divided by the temperature. The Second Law reveals a

---

*William Thompson (1824–1907) was knighted in 1866 and granted a peerage, as Baron Kelvin of Largs, in 1892. He is generally known as Kelvin or Lord Kelvin, although the peerage was conferred long after his important achievements in science. He will be referred to as Kelvin or Lord Kelvin throughout this book.

fundamental lack of symmetry in Nature: You can trade work one for one for heat, at the exchange rate of 4.18 joules per calorie,* but you cannot trade heat one for one for work. There is a tax added for work. That tax is entropy—more precisely, increase in entropy, and it is due to the tendency of thermal energy to disperse every which way. Heat can produce coherent motion that results in work, but simultaneously it must also produce incoherent, chaotic motion, which cannot be recovered as useful work.

The absence of symmetry expressed by the Second Law gives direction to Time. Time flies in the direction of increased disorder or entropy, and that was Walter Alvarez's point. Until that evening I had not considered the deeper meaning of the Second Law because as a student I had only learned to apply the law, in a rather prosaic and mechanical fashion, to test the degree of equilibration among minerals in rocks.

Because the Second Law defines equilibrium in terms of changes in the internal energy of a system, it also provides ways to relate the effects on a system, such as a rock, of changes in temperature, pressure, volume, or chemical composition. We need go no deeper into the subject than to know that natural systems seek conditions of chemical and physical equilibrium, that conditions of equilibrium change with changes in temperature, pressure, volume, and chemical environment, and that these relationships may be quantified and reproduced. In other words, they are scientific facts.

The earth and planetary sciences have benefited from the discipline and structure introduced by chemists and physicists who developed concepts such as the Laws of Thermodynamics through theory and experiments with simple systems. However, the application of data and concepts derived from simple systems to the complex world of natural rocks was not always welcomed by the geologists. Natural tensions arose between those who understood manifestations of natural laws in laboratory experiments and those

---

*One calorie (1 cal) is the quantity of heat that must be transferred to one gram of water in order to raise its temperature one degree centigrade. Dieters note: The "Calories" you count are really kilocalories (1,000 cal). One joule (1 J) is the amount of work accomplished when a force of one newton is displaced one meter in the direction of the force. One newton is the force that accelerates a mass of 1 kilogram 1 meter per second per second.

who, while never denying the authority of natural laws, were more concerned with the complexity of the natural world and less inclined to trust understanding derived from simple systems. The tension is apparent in a letter, written during the late 1940s, by a prominent experimental geologist to an equally prominent field geologist:

> Having had occasion some years ago to learn the art of lip reading, I noticed yesterday when I was giving my paper that at the end of each of my sentences you said, "Horse Shit." Evidently you had made special note of the word equilibria in the title of my paper and were from time to time reminding yourself and your neighbours of the gist of the discussion. You are, however, under a misapprehension as to the derivation of the word equilibria. It does not come from *equus* = a horse and *libria* = things liberated or discharged, but is from quite different roots. If you will consult a chemist you will be able to learn the real significance of the word and I may add that I feel that one so highly placed in geological circles as you should make it a point to acquire some familiarity with the exact significance of common terms in collateral sciences.[26]

Atoms and fluids diffuse or flow so slowly through or around crystals and other solid bits of rocks that rocks seldom reach chemical equilibrium. This failure to equilibrate completely is why rocks are so interesting. Every rock contains a more or less complete record of every chemical and physical change that befell it. Therefore, it is like a history book with some, but not all, of the pages torn out or smudged. Usually the story is preserved in the text, if only we are clever enough to reconstruct it.

Three

Hot Enough to Melt

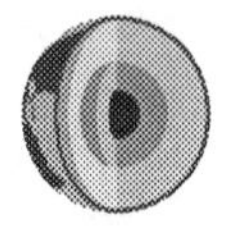

That the Earth is hot inside might be among the most ancient of human insights. The thought of heat within the Earth must have occurred to many who saw volcanic eruptions, who bathed in hot springs, or who worshipped at burning tar pits, yet the importance of internal heat was not generally recognized until the eighteenth century. In Europe, Aristotle's (384–322 B.C.) notion that sunshine and starlight produce ore-forming exhalations and other signs of heat went unchallenged for nearly two thousand years. Of course, there were iconoclasts, such as Strabo, a Greek geographer and historian at the time of Christ, who attributed earthquakes, mountains, and volcanoes to central fires.

The central fires flared in the Age of Faith. The visions of Saint Isadore (A.D. 560–636), seventh-century archbishop of Seville, included two rather peculiar cone-shaped furnaces within the Earth to cook sinners. Then came Dante, who put an inferno down below in 1200 but hedged his bet by freezing the deepest levels of the underworld.

Physical evidence joined the supernatural during the Renaissance as miners, scholars, and alchemists began to dismantle the imaginary world of Greek philosophers and Christian saints, but science was still far off. Georges Agricola, fifteenth-century physician, mining geologist, and metallurgist, claimed to disregard "all those things which I have not myself seen, or have not read or heard of from persons upon whom I can rely."[27] Yet he wrote of fires in the

Earth that consumed mountains. (Perhaps he was referring to burning coal seams, which certainly occur.) In the following century, Thuringian Jesuit and professor Athanasius Kircher imagined Earth with a blazing core, a mantle with many chambers of fire, chambers of molten rock, and conduits connecting these chambers with each other and with volcanoes on the surface. And in the middle of the seventeenth century, René Descartes proposed an earth model of spherical shells of gases, water, rocks, and metal upon a core of incandescent, sun-like fire. In spite of their grip on the imagination, the central fires were snuffed out in the 1770s when Karl Wilhelm Scheele, Antoine Lavoisier, and Joseph Priestley discovered oxidation and the nature of combustion. In the absence of air, deep subterranean fires would soon go out.

Even so, the planet was obviously hot: Its roundness was taken to mean that it had once been molten; temperatures increased with depth in wells and mineshafts; and volcanoes were widespread. For the next hundred years, geologists and physicists struggled with data that were increasingly discordant. Where was the heat coming from? Was the planet simply cooling down from a white hot liquid? If so, the planet was still young enough and hot enough to produce molten lavas. Given such obvious evidence for internal heat, Kelvin calculated Earth's age to be between 20 million and 400 million years. His uncertainty was due largely to ignorance of thermal conductivities, but he was more comfortable with the younger age. And what about the Sun? If, as the physicists believed, the Sun's heat derived from gravitational collapse, then after only a few hundred million years the Sun would be dead, and Earth's surface would be icy cold—too cold for liquid water or life. Rocks and fossils, however, demanded vastly more time than a few hundred million years. Physics and geology were deadlocked, but in 1896 Henri Becquerel (1852–1908) provided the key to breaking the dead lock by discovering radioactivity.

Much has been said of serendipity in scientific discovery, yet when accounts of scientific discovery are examined, one often finds a long trail of conscious or unconscious preparation. In the case of Henri Becquerel's discovery of radioactivity in

1896, the trail extends back three generations. Both Henri's grandfather, Antoine Charles Becquerel, and his father, Edmond Becquerel, were physicists who studied electrical and magnetic phenomena. Edmond devoted much of his attention to substances that glow in the dark, such as uranium ores, distinguishing for the first time those that are phosphorescent from those that are merely fluorescent.*

By the late nineteenth century, when Henri was at work, physics was sparkling in the glow of new forms of radiation, such as Röntgen's mysterious X rays. It is thus not surprising that Henri paid more than casual attention to the unexpected fogging of photographic plates that had been wrapped in light-tight paper and stored in a drawer with some uranium salts. His discovery of a penetrative radiation that did not depend upon external excitation was a surprise, but it was no accident.

Thanks to the ideas and experiments of Henri Becquerel, Marie and Pierre Curie, Ernest Rutherford (1871–1937), and a continuing succession of scientists, we now understand what fogged Henri's film. Radiation, with enough energy to pass through the film wrapper and expose the film, was coming from the uranium in the drawer. Over the next 15 years, Rutherford and others determined that there were actually three kinds of radiation; two were emissions of high-energy particles, and the third was a kind of electromagnetic wave, like light and X rays. All were associated with the atomic nucleus, which Rutherford also discovered.

Atoms have massive nuclei built of protons and neutrons. All atoms of the same element have the same number of protons, which defines the atomic number, but not all of them contain the same number of neutrons. As the number of neutrons varies, so varies the atomic mass. Atoms of the same element with different masses are referred to as different *isotopes* of the element. Certain isotopes are unstable, which means that their nuclei show a tendency to disintegrate, or decay, over some period of time in a spontaneous process, called *radioactive decay,* which occurs in Nature as

---

*Both phosphorescent and fluorescent substances glow because they have absorbed energy, such as sunlight. Fluorescent substances stop glowing when the energy supply is cut off; phosphorescent substances continue to glow for some time after the energy supply is cut off.

well as in the laboratory. We express the tendency to decay in terms of the *half life* of the isotope, the time required to reduce the number of unstable nuclei by half. The rates at which different isotopes decay are, so far as we know, constant and unaffected by time, pressure, temperature, or chemical environment. These decaying isotopes can be used to measure the absolute age of a rock or a crystal, but here our concern is with heat generated by radioactive decay.

Isotope	*Name*	*Half Life*
^{14}C	Carbon 14	5.57×10^{3} yr
^{40}K	Potassium 40	1.28×10^{9} yr
^{87}Rb	Rubidium 87	4.7×10^{10} yr
^{147}Sm	Samarium 146	1.06×10^{11} yr
^{238}U	Uranium 238	4.51×10^{9} yr

Nuclear disintegration involves the emission of two kinds of particles from the nucleus: alpha particles, consisting of two protons and two neutrons, and beta particles, which are identical to electrons. The nucleus also emits high-energy, electromagnetic radiation, called *gamma rays*.

The heat production associated with radioactivity was discovered early. Marie Curie (1867–1934)was a newly married doctoral candidate at the Sorbonne when Becquerel discovered radioactivity. As her doctoral research she chose to search for other radioactive elements in addition to uranium. She discovered polonium and radium. In 1903 she completed her thesis, *Recherches sur les substances radioactives*, and shared the Nobel Prize in Physics with her husband Pierre and Henri Becquerel. That same year Pierre Curie (1859–1906) and his assistant, Albert Laborde, published a report, "On the Heat Spontaneously Disengaged by Radium Salts."

Spontaneous nuclear disintegrations are accompanied by conversion of tiny amounts of mass to energy. Although the quantities of mass are minute, the quantities of energy that result are great. A gram of the highly radioactive element radium, which is about the size of a pencil eraser, gives off nearly two calories of heat per second—about the same as the output from a small Christmas tree light bulb.*

---

*Two calories per second equal 8.37 watts.

Einstein's famous equation, $E = mc^2$, relates energy (E), mass (m), and the velocity of light (c). The mass of a proton or a neutron is equivalent to 931 million electron volts* (3.6×10^{-9} calories), which may not seem like much until you consider the numbers of protons and neutrons. A block of granite weighing 12½ tons, which could be a 5-foot cube, contains approximately 40 grams of radioactive potassium. The decay of 40 grams of radioactive potassium, which could be a 1½-inch cube, releases as much energy as the explosion of 32 tons of TNT! Of course, nuclear disintegrations in rocks involve only tiny mass conversions and generally go slowly, according to the half life of the particular isotope. But go they do. Throughout the Earth, more so in the continental crust than in the oceanic crust or deeper portions of the planet, unstable nuclei are constantly breaking up, heating the planet as they disintegrate.

Radiogenic Heat Production[28]

Granite (continental crust)	7.2×10^{-6} cal/yr/g
Basalt (oceanic crust)	1.2×10^{-6} cal/yr/g
Peridotite (mantle)	2.4×10^{-8} cal/yr/g

Rocks such as granites contain enough heat-producing radioactive atoms to dominate the Earth's heating system. As R. J. Strutt (1842–1919), who later became Lord Rayleigh, pointed out, the radioactive elements must be concentrated near Earth's surface or our planet would be hotter than it is.[29] Strutt's insight was important because it showed that the crust differs chemically as well as physically from the interior.

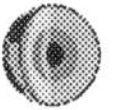

Volcanic eruptions, hot springs, and geothermal wells are obvious indications of Earth's internal energy, and there is less obvious but equally important evidence. Mountain

*Not to be confused with a volt (a unit of electromotive force or electrical potential difference), an electron volt is a unit of energy. One electron volt (eV) is the energy acquired by any charged particle, such as an electron, carrying a single electronic charge, when it falls through a potential difference of one volt.

ranges, for example, represent colossal concentrations of chemical, gravitational, and strain energy. This is Earth's internal energy, which also powers earthquakes.

We can draw a diagram showing the probable range of temperatures that occur within the Earth. This curve is controlled by a few facts. We know, from the transmission patterns of earthquake waves through the planet, where the Earth is mostly or entirely solid and where it is mostly or entirely molten; we can make assumptions, which are at least self-consistent, about the compositions of various portions of the planet; and we have rocks that actually record temperatures and pressures from the outer 450 kilometers of the planet.

The rocks that record temperatures and pressures from as deep as 450 kilometers were brought to the surface very quickly in explosive eruptions. They erupted so quickly that the minerals in them had no time to adjust to lower temperatures and pressures. About twenty years ago, an experimentalist named F. R. Boyd and his colleagues discovered that the distribution of calcium, magnesium,

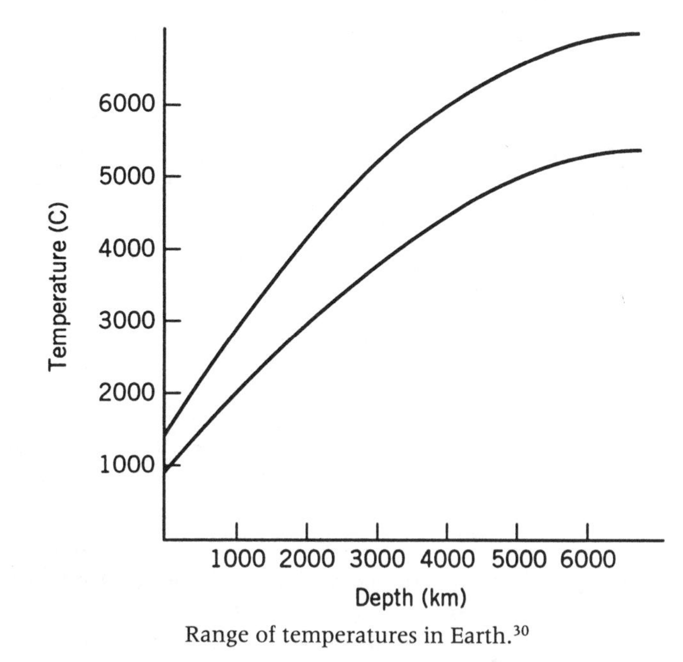

Range of temperatures in Earth.[30]

iron, and aluminum between coexisting garnet and pyroxene minerals in these rocks is different at different pressures and temperatures. By measuring the distribution of these elements between the coexisting minerals, Boyd and others were able to determine the pressure and temperature at which the garnets and pyroxenes came to chemical equilibrium when they lay deep in the Earth.

On the average, 1.5 microcalories (1.5×10^{-6} calories) of heat flow to each square centimeter on the Earth's surface every second, which amounts to 2×10^{20} calories of heat flowing out from the entire planet over a year's time. This seems like a great deal, and given time and some means to concentrate the energy it is sufficient to move mountains or melt rocks. However, it is no more than 0.02 percent of the solar energy received by Earth. Consider that the 1.5 microcalories per square centimeter per second flowing from Earth's interior would take more than twenty months to melt a centimeter of snow.

Heat moves by conduction (each atom jostling its neighbors), convection (hot materials moving to displace cooler ones), or radiation (glowing). Rocks are very poor conductors of heat, and, at the very high pressures of Earth's interior they may be opaque to the infrared radiation that transmits heat. Thus, it appears that within the Earth most heat is moved not by radiation or conduction but by convection—gravity moving the rocks themselves, or fluids, according to their relative densities, from one place to another. Convective processes not only are responsible for transporting heat, but are also responsible for concentrating heat so that it can accomplish great works such as earthquakes, mountain belts, or volcanic eruptions.

Kilimanjaro, Taal, Mayon, Fujiyama, Ararat, Santorini, Etna, Stromboli, Vesuvio, Popocatepetl, Kilauea, St. Helens, Pinatubo—active volcanoes have accompanied us as our civilizations have risen and declined, so we have always known that some rocks come from molten lavas. Until the eighteenth century, however, such melting was presumed to be a local phenomenon, caused by subterranean fires. This was not an unreasonable idea, since coal seams can and do ignite and burn with great intensity for years. However, the idea failed when early geologists recognized enormous quantities of igneous rock in the Earth's crust, but this failure came only after one of the most intense debates in the history of science.

The first indication that igneous rocks might be found far from active volcanoes (both in distance and in time) came in 1752. That was when French naturalist Jean Etienne Guettard (1715–1786), gathering data for the first geological map of France, came upon the volcanoes of the Auvergne, cold for 20 million years but looking as though they had erupted yesterday. A decade later another French naturalist, Nicolas Desmarest (1725–1805) also visited the Auvergne. Impressed by the flows of basaltic lava and, in particular, by the development of columnar jointing in them, he concluded that basalts everywhere are frozen lavas.

When hot solid lava flows or thin intrusive sheets cool, they shrink and break into straight-sided columns with simple polygonal—often hexagonal—cross-sections. These columns typically extend vertically through the flow from the surface to the base. There are many places to see this phenomenon, known as columnar jointing, in basalt and other volcanic rocks. Famous displays are found at Devil's Postpile National Monument, near Mammoth Lakes, California, the Columbia River region of Washington and Oregon, Fingal's Cave in Scotland, the Giant's Causeway in Ireland, and the Palisades of the Hudson, near New York City.

Desmarest's ideas caused great excitement as other observers noticed basalts and columnar jointing in places where volcanism had never been suspected. The study of igneous rocks seemed to be off to a fast start. Then, from his lectern at the Academy of Freiberg in 1787, a great and persuasive Professor of Mineralogy, Abraham Gottlob Werner (1749–1817), threw cold water on the whole idea.

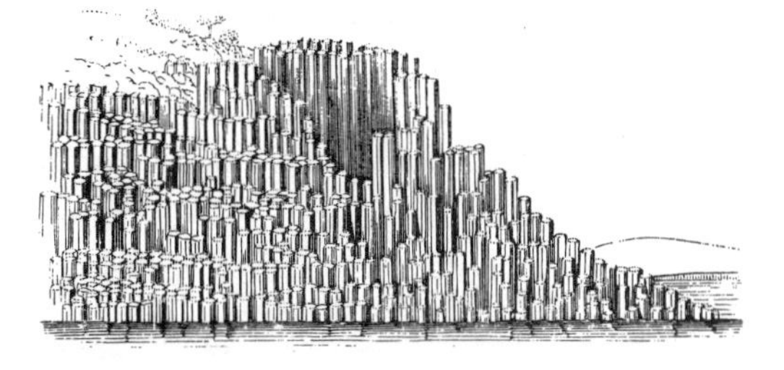

Columnar basalt.[31]

"I hold," said he, "that no basalt is volcanic, but that all of these rocks . . . are of aqueous origin."[32]

According to Werner and his students, the world had been enveloped by a great ocean, and from this ocean all crustal rocks, from granite to limestone, with very few local exceptions, had been deposited. By all accounts Werner was an inspiring teacher, devoted to his mineral collections, his students, and his point of view. Consequently, geologists were divided into two camps: Werner's, called neptunists, and Desmarest's, called vulcanists. Eventually the vulcanists won, not because they were better or smarter, but because they were correct. Many of Werner's students, such as Jean François D'Aubuisson de Voissins (1769–1819) and Christian Leopold von Buch (1774–1853), visited the Auvergne and other volcanic terrains and saw for themselves the evidence for the igneous origin of basalt.

The evidence of the Auvergne was incontrovertible and compelling. D'Aubuisson faced the facts without flinching:

> The facts which I saw spoke too plainly too be mistaken; the truth revealed itself too clearly before my eyes, so that I must either have absolutely refused the testimony of my senses in not seeing the truth, or that of my conscience in not making it known. There can be no question that basalts of volcanic origin occur in Auvergne . . .[33]

Von Buch, who visited the Auvergne before D'Aubuisson, was equally impressed but less able to immediately abandon the teachings of his beloved professor. That took more time and more observations. Von Buch was a fine man whose memory might serve as a fitting monument to all the intrepid geological explorers of his time. Another great geologist, Sir Archibald Geikie (1835–1924), described him with affection:

> Among the leaders of geology in the first half of this [nineteenth] century there was no figure more familiar all over Europe than that of Von Buch. Living as a bachelor, with no ties of home to restrain him, he would start off from Berlin, make an excursion to perhaps a distant district or foreign country, for the determination of some geological

> point that interested him, and return, without his friends knowing anything of his movements. He made most of his journeys on foot, and must have been a picturesque object as he trudged along, stick in hand. He wore knee-breeches and shoes, and the huge pockets of his overcoat were usually crammed with note-books, maps, and geological implements. His luggage, even when he came as far as England, consisted only of a small baize bag, which held a clean shirt and silk stockings. Few would have supposed that the odd personage thus accounted was one of the greatest men of science of his time, an honoured and welcome guest in every learned society of Europe.[34]

Today we know basalt to be the most common lava on the face of Earth, the Moon, and the terrestrial planets. Most of us have seen it as the black lavas of the Columbia and Snake River gorges, of Hawaii and Iceland, of the Palisade cliff along the Hudson, or of the dark maria (seas) of the lunar surface. Because it freezes rapidly on or near the Earth's surface, basalt is a fine-grained rock, so fine-grained that one needs a magnifying glass to discern the individual crystals. With a glass, or better a microscope, one sees that it is dominated by two common silicate minerals, in roughly equal proportions: plagioclase feldspar and pyroxene. Both are rich in alkali metals, aluminum, calcium, iron, magnesium, and silicon. A variety of less abundant minerals, also occur in basalts, increasing their diversity. Directly or indirectly many of the mineral resources of Earth's crust derive from these abundant rocks. When molten basalt freezes slowly, well below the surface, it forms a coarse-grained igneous rock called gabbro.

The origins of igneous rocks that freeze beneath Earth's surface—the intrusive rocks such as gabbro—are more obscure than the origins of lavas. An erupting volcano is persuasive evidence, but no one has seen intrusive rocks crystallize. Understanding their origin required a major intellectual leap, and the genius who made it was a Scottish physician and farmer, James Hutton (1726–1797), a vulcanist who is today regarded as the founder of modern geology. Hutton took Desmarest's igneous argument a giant step forward by suggesting that not only basaltic lavas had formed from molten material (magma) but also mountainous masses of granite. These coarser-grained rocks must have frozen slowly, deep beneath the Earth's surface. There was no possibility of observing the process directly because no volcanic lavas freeze into granite. Nor was there any possibility of demonstrating the process in the laboratory because no eighteenth-century experimental apparatus could generate the pressures and temperatures required to form granite.

Hutton's idea was just that, an hypothesis based on field evidence. Field evidence may be persuasive or it may not, as different minds may find different messages in the rock outcrop. Not only Werner and the neptunists but Desmarest himself thought Hutton wrong about granite. This "Granite Controversy," begun by Hutton, lasted nearly two centuries until enough field, experimental, and chemical data had been gathered to resolve the dispute by showing that "there are granites, . . . and there are granites."[35] or, in other words, some granites are igneous (probably) and others (perhaps) are not.

Hutton's contributions to earth science were not merely the magmatic model for granites. His *Theory of the Earth*, presented in 1785, contained an extraordinary number of fundamental geologic concepts: *The Principle of Uniformitarianism*—that "the present is the key to the past," or that all features of the geologic record can be interpreted and understood in terms of processes that are going on today; *Metamorphism*—that one rock can be changed to another under conditions of increased heat and pressure; *Plutonism*—that granites and certain other coarse-grained rocks are igneous in origin, crystallizing far beneath the surface; *Tectonic Uplift*—that sedimentary beds may be pushed up by Earth's internal energy to form mountains; and, above all, *The Rock Cycle*—that all rocks are related to one another through processes of erosion, sedimentation, metamorphism, and magmatism. The unavoidable conclusion of Hutton's Theory of the Earth was that Earth must be very, very old. "The result, therefore, of our present enquiry," he wrote, "is, that we find no vestige of a beginning—no prospect of an end."[36] This was, of course, a direct afront to the Church, and in another century he would have been broken on the rack and burned alive. More important, Hutton provided an unprecedented concept of time that would lead to the Theory of Evolution and beyond.

The igneous processes that the vulcanists discovered could not be understood until well into the twentieth century, when experimental and analytical techniques would be available to study processes and products occurring deep within the Earth. These processes involve a substance known as *magma*, a naturally occurring, complex material consisting of one or more molten rock liquids, with or without suspended crystals (called *phenocrysts*) of one or more minerals, and with or without bubbles of a separate gas phase. All igneous rocks form from magmas. Until very recently most magmas were not susceptible to direct observation unless they happened to erupt as lavas. The existence of most magmas is based upon structural and textural observations of cold rocks; chemical evidence for their existence is seldom compelling; so we should acknowledge that our subject remains, to some extent, hypothetical.

The most important part of most magmas is the silicate melt. Molten rocks do not lend themselves to convenient observation either in the field or in the laboratory. We can study glasses, which are quenched (quickly frozen) melts, but direct observations of melts themselves are more difficult. Nevertheless, what we've seen of its electrical conductivity, viscosity, molecular vibrations, and X ray diffraction tell us that much of the melt is ionized—more so as temperature rises; partially ordered—on a local scale; and more or less polymerized (interconnected)—more so with increasing silica content. Observations of this kind, made in the laboratory, seem to explain some natural behavior of magmas: why they do or do not erupt, why they tend to be explosive or calm, and why they form gently sloping mounds or steep-sided cones.

About 1825 an experimenter named Drée discovered that when a basaltic* melt freezes it does so over a range of temperatures and that the first crystals are likely to sink to the bottom of the crucible before the entire mass solidifies. Field evidence for this phenomenon

* The composition of the melt is assumed to have been basaltic. Drée did not publish his results, and those who referred to them described the material simply as "lava." From the rapid crystal settling that Drée observed, however, it is likely that he was working with a low viscosity melt such as molten basalt.

was described from lava flows on the Canary Islands by Werner's student, Leopold Von Buch, and later by the young Charles Darwin (1809–1882). Sixty-two years after Drée's experiments, a Canadian was born who was to become the preeminent petrologist (rock scientist) of the twentieth century by the systematic, quantitative study of this process, called *fractional crystallization*. By melting and freezing simple combinations of elements that represented rocks or their constituent minerals, Norman Levi Bowen (1887–1956) created a model for explaining the diversity of igneous rocks. His model began with basalt as the parent liquid and derived various other magmas by extracting crystals of different compounds from the basaltic liquid.

Fractional crystallization is a process that separates crystals from their parent liquid. When crystals grow in a solution containing many chemical elements, they extract only those elements that they can accommodate and leave the rest in the solution. The solution, therefore, becomes enriched in those elements that are left behind. Imagine a glass of salt water. If you place the glass in your freezer, ice crystals will begin to form. What will happen to the remaining salt water? If you taste it from time to time, you will find that it becomes progressively saltier as the ice crystals grow because pure water is being extracted to form the ice. The same processes of fractionation and enrichment occur in magmas.

Normally crystals and melts have different densities. Most early-formed, high-temperature minerals, such as olivine, pyroxene, and calcium-rich plagioclase, are more dense than the melt, and so they sink. Some igneous rocks are actually magmatic sediments; that is, they are composed of accumulations of crystals that settled out of their magmas. Such accumulations can be valuable. Magmatic accumulations of chromite ($FeCr_2O_4$), for example, are the world's main source of chromium metal.

On a grander scale, the Earth, or a portion of it, may have differentiated by fractional crystallization of melts during core formation. The Moon was apparently differentiated by fractional crystallization that involved crystal settling and flotation. The light-colored lunar highlands may have formed by flotation of plagioclase crystals on a magma ocean that enveloped the entire Moon.

The most abundant elements in common terrestrial magmas are oxygen, silicon, aluminum, titanium, iron, manganese, magnesium, calcium, sodium, and potassium. Fractional crystallization in these magmas often involves early-formed oxide or silicate crystals rich in calcium, magnesium, or iron, and the residual liquids are often enriched in alkali metals and other elements the early-formed crystals left behind. Such chemical trends are common among igneous rocks and can be observed on scales ranging from microscopic to mountainous. The attribute of fractional crystallization that caught Bowen's fancy, and that supports his ideas about the parental relationships among igneous rocks, is the enormous range of liquid compositions that are generated during one continuous crystallization episode. If physical separation were achieved at various times during the crystallization episode, many different derivative melts could come from one source.

Bowen's influence over his own and succeeding generations of geologists was extraordinarily strong for several reasons. He *was* brilliant, creative, and energetic, but others had those gifts and were far less influential. Bowen was special because he was working in the fertile, unplowed ground between experimental physical chemistry and geology with self-confidence, wit, and grace of expression. He applied new insights and techniques to traditional problems and defined fruitful new areas for investigation. As evidence of his self-confidence, here is Bowen responding to the first question from his Ph.D. thesis defense: "Gentlemen, do you wish me to give the answer I know you want, or do you wish me to give what I think is the right answer?"[37] As an example of wit and grace, here is Bowen speaking towards the end of his career:

> Magmas are the stuff of which the igneous rocks are born. It would be splendid if I could here and now proceed briefly to tell you all about magmas. That, of course, cannot be done, but it would not be unreasonable to expect that one who has spent the better part of a rather long life in the study of magmas could perhaps summarize what is known about them. Unfortunately, statements of what is known must of necessity be so wound about "with circumstance" that it is much easier to emphasize what is not known. There one can be, for the most part, thoroughly dogmatic. Indeed, contemplation of our knowledge of magmas is not likely to put one in the mood of the lassie whose heart to her "mou' gied a sten." Not delight but

> dispair might readily fill our beings were it not for the fact that in the scientific lexicon there is no such word as despair. We are never devoid of some hope that we shall eventually know everything. It may often be salutary, nevertheless, to recognize the remoteness of that time and to take stock of our ignorance.[38]

Magmas form when rocks melt. However, because most rocks contain more than one mineral, the melting process is not simply a change from solid to liquid; it is a chemical reaction involving the newly formed liquid and the different minerals. The simplest melting reaction is called *eutectic melting,* and it occurs between many minerals, such as the pyroxene called diopside and the calcium feldspar called anorthite. Diopside melts at 1,391°C, and anorthite melts at 1,550°C, but *any* mixture of diopside and anorthite crystals will begin melting at 1,270°C, the eutectic temperature. Moreover, the liquid that appears at 1,270°C always has the same chemical composition.* Whether the melting mixture starts with 99 percent anorthite crystals and 1 percent diopside crystals or vice versa, the first liquid to form will be 42 percent molten diopside and 58 percent molten anorthite, dissolved together in one homogeneous, very hot solution. This is an essential characteristic of eutectic systems: All combinations of the crystalline components will produce identical liquids when they melt together.

Eutectic melting may not be familiar, but many people use it unknowingly. An example is ice mixed with common salt, which has a eutectic temperature of –21°C (–4°F). When we sprinkle salt crystals on an icy sidewalk at any temperature above –21°C we induce eutectic melting, the salt and the ice react to form brine, which flows into the gutter. In a tastier application, the mixture of ice, salt, and brine chills an ice cream maker to the eutectic temperature. Another familiar example of a eutectic system is automobile antifreeze. Antifreeze is a solution of water and ethylene glycol with the eutectic temperature of –60°C. The addition of glycol to water

---

*At 1 bar of pressure. The bar is a unit of pressure that is very nearly equal to atmospheric pressure.

dramatically lowers the temperature at which ice crystals begin to form, so adding 30 percent glycol, for example, prevents any ice from forming above –20°C.

The most abundant magma of the Earth, the Moon, and the terrestrial planets, in both space and time, is basaltic. Why? Of all the processes we know, eutectic melting is the one most likely to produce the same melt composition, over and over and over again, from rocks that contain the same major minerals irrespective of their proportions. The mantle rocks of Earth and the terrestrial planets are not likely to behave as ideal eutectic systems, but they may well contain similar mixtures of minerals that melt in a eutectic-like manner.

The high-temperature experiments by Norman Bowen and others explained much about the behavior of silicate magmas at or near Earth's surface, but magmas are born and evolve far below the surface, under pressures that far exceed atmospheric pressure (1 bar). How would silicate rocks and magmas behave under high pressure? How would they behave in the presence of volatile substances, such as water or carbon dioxide, which could not be included in unconfined, low-pressure experiments?

The effect of pressure on the melting temperature of simple systems can be estimated indirectly with a thermodynamic relationship called the *Clapeyron Equation,** which relates this effect to the change in volume during melting and the quantity of heat required to melt the substance. Experiments to determine the change of volume and heat of fusion at atmospheric pressure were accomplished in the 1890s by a scientist at the U.S. Geological Survey named Carl Barus.[39] From his results curves could be drawn showing the melting temperature of dry rocks increasing with increasing pressure, but direct studies of high-pressure melting were delayed by technical problems. Although melting experiments could be made with silicates in open vessels on a lab bench as early as 1727, techniques for making comparable experiments at high pressure were not perfected until around World War II.

---

*The Clapeyron Equation: $\Delta T/\Delta P = (T\Delta V)/\Delta H$, where the symbol Δ indicates finite changes of, for example, temperature (ΔT), and ΔH stands for the heat of fusion, which is the quantity of heat required to convert a solid at its melting temperature to liquid at the same temperature.

To study the realm of the Earth's outer crust one needs a device that can compress a sample with a force of at least 5,000 bars (5 kilobars), heat it to 500°C, and simultaneously measure and control the pressure and temperature. Furthermore, one must be able to *quench* the experiment, that is, bring the sample rapidly back to room temperature in order to preserve it as it was at high temperature and pressure. One technical solution is the so-called cold-seal bomb, a hollow steel cylinder filled with pressurized water and heated (except for the cold valve, or seal) within a furnace. Bowen and others used cold-seal bombs to investigate chemical relationships between minerals such as feldspar and mica, and some low-temperature magmatic situations, but the origin of basaltic magma lay deeper in the Earth, beyond the cold-seal bomb's range.

With advances in metallurgy and engineering, the pressures and temperatures attainable in laboratory experiments crept upwards. By 1938 melting experiments had been performed at 4 kilobars and 1,200 degrees. World War II interrupted this progress, although it provided novel materials and techniques. In the early 1950s, by compressing argon gas in an internally heated steel cylinder, experimental geologist Hatten Yoder determined the melting curve of a major component of basalt, the pyroxene diopside, up to 5 kilobars. During the next several years, Yoder and his colleagues extended their experimental range to 10 kilobars, which corresponds to Earth depths of 30 to 35 kilometers.

Still higher pressures and temperatures were attained by F. R. Boyd and Joseph England with an internally heated apparatus that employed a solid medium such as salt or talc rather than argon gas. This apparatus extended the range of experiments to 40 kilobars and 1,600°C. By the mid-1960s experiments were being performed under conditions that obtain at the depths of basaltic origins—100 or more kilometers below the Earth's surface.

Among the experimental results was confirmation of the prediction that the melting temperature of a dry rock increases with increasing pressure. This means that a hot dry rock, deep in the Earth, can be melted by being raised toward the surface and thereby reducing the pressure. Most basaltic magma may be generated by this mechanism. The relationship between melting temperature and pressure was not surprising. Even on a lab bench in the open air one

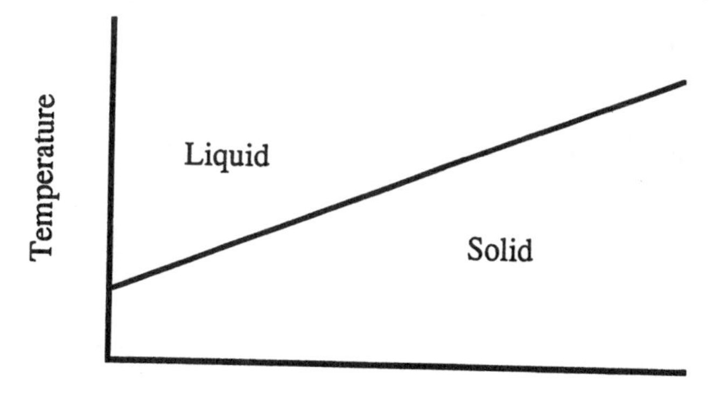

could observe that the melt occupied more space than the rock from which it formed, so it seemed likely that higher temperatures would be required to melt dry rocks at high pressure. More surprising were the results of melting wet rocks at high pressure. In this case, the temperature–pressure relationship, was reversed. A wet rock—in other words, a rock coexisting with supercritical H_2O, not liquid water—melts at decreasing temperature with increasing pressure! Thus, a hot, wet rock could be melted by squeezing it harder, and a wet melt could be frozen by reducing the pressure—by eruption from a volcano, for example.

Just as explorers and voyageurs used rivers to penetrate and learn about the American hinterland, so can we learn some things about the Earth's upper mantle by studying volcanoes. Volcanic gases and melts provide information about the chemistry of the source area. Violent eruptions carry solid fragments out from the interior. These fragments may tell us about the compositions, structures, temperatures, and pressures of the rocks through which the magmas passed, or they may tell of crystal fractionation processes within the magmas themselves. More fundamental still, the presence of a volcano tells us that conditions in the underlying mantle are (or were) appropriate for melting.

Between depths of 100 and 250 kilometers the Earth's thermal gradient and the melting curve of the presumed mantle rocks are very close and may, in places, overlap. The Earth appears to be relatively soft and malleable in this region, and seismic waves are slower here, probably because of the presence of small amounts of interstitial melt. Most sustained magma generation probably involves decompression of materials from this low-velocity zone rising towards the surface. The cause of upward movement may be rifting (tearing apart) of the overlying lithosphere or some convective motion in the mantle.

Most volcanism is located along so-called "accreting" or "consuming" plate margins, where the lithosphere, Earth's 100-kilometer-thick outer shell of relatively cold, rigid rocks, is created or destroyed. The accreting margins are the mid-ocean ridges where oceanic lithosphere is made. We can roughly estimate the quantities of magma and the rates of magma production there as follows:

1. The oceanic crust contains about 7½ kilometers of magmatic rocks (lavas and underlying intrusive rocks), and the area of the oceanic crust is approximately 360 million square kilometers. Therefore, the volume of this magmatic material is 2,700 million cubic kilometers:

$$7.5 \text{ km} \times 3.6\times10^8 \text{ km}^2 = 2.7\times10^9 \text{ km}^3.$$

2. The oldest oceanic crust in the ocean basins today is 200 million years old, so the annual production of magma along the mid-ocean ridge system is about 14 cubic kilometers per year:

$$2.7\times10^9 \text{ km}^3 \div 2\times10^8 \text{ yr} \cong 14 \text{ km}^3 \text{ per year.}$$

3. The area of crust produced is about 1.9 square kilometers per year:

$$14 \text{ km}^3 \div 7.5 \text{ km} \cong 1.9 \text{ km}^2 \text{ per year.}$$

4. The mid-ocean ridge system is 60,000 kilometers long. Spread along the entire ridge system, this annual areal crust increase

amounts to an average annual spreading rate of 3.2 centimeters, which is about what we observe.

$$1.9 \text{ km}^2 \div 6\times10^4 \text{ km} = 3.2\times10^{-5} \text{ km or } 3.2 \text{ cm}$$

As the lithosphere is pulled apart, probably by gravitational forces, partially molten material moves up to fill the opening space. The magma does not push the plates apart. Why not? Because it does not contain enough excess internal energy. Cold steam won't drive a piston; it condenses instead. Similarly, if the internal energy of the rising magma were required to push the lithosphere, the magma would freeze.

The lavas that come out on the sea floor are basaltic. Their temperature is 1,100 to 1,200°C, but when they hit the water their outer surfaces freeze in seconds.* Signs of rapid quenching, such as glass and feather-like quench crystals, are common. Characteristic forms are "pillows," bladders of quenched lava containing several gallons of still-molten magma. As lava is extruded through cracks in the erupting mass, it is immediately covered by a skin of glass; pillows grow by constantly dilating, cracking, and requenching this glassy skin. When they reach some critical size, which depends upon the viscosity of the magma and the steepness of the slope, they break loose and tumble to rest in languorous heaps on the sea floor. We find pillow lavas in the geologic record way back into the Precambian Era, at least 3,500 million years ago.

In some places along the mid-ocean ridges the volume of magma generated is much larger than normal, over a significant period of time. At such places volcanic piles may rise above sea level and form islands, such as Iceland. Normally, however, the production of magma along the ridges appears to have been remarkably uniform over geologic time. Spreading rates recorded in the existing oceanic crust range between 1 and 20 centimeters per year.

---

*The water is cold (0°C), but the extraordinarily rapid quenching is due to the fact that water has a high heat of vaporization (540 cal/g) and a very high heat capacity (1 cal/g°C).

The process of basalt generation along the mid-ocean ridges can be diagrammed like this:

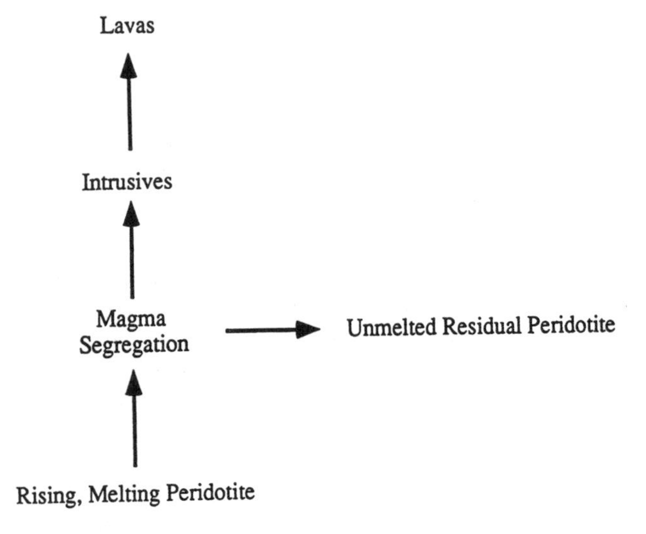

We can never observe this process directly; however, chemical, physical, and geological facts *suggest* that the mantle material that rises to partially melt and yield basaltic magma is a rock called peridotite, named for peridot, the apple-green gem variety of olivine, which is the birthstone for August. Olivine constitutes 60 percent or more of peridotite.* Rare though peridot is, ordinary olivine is probably one of the most abundant silicates in the Earth, for it apparently dominates the mantle down to about 400 kilometers, which is 15 percent of Earth's volume. The other minerals in peridotite are pyroxenes, such as diopside, and, depending upon the depth of crystallization, small amounts of either feldspar (shallow), spinel (intermediate), or garnet (deep). Because of its deep and mysterious origins, garnet-bearing peridotite excites passionate enthusiasm in

---

*Many rock names are derived from the name of the rock's most prominent mineral by adding the suffix -ite to the mineral name (e.g., peridot → peridotite).

some geologists. Other people experience rather different reactions. Take my wife Lynn:

> What's a garnet peridotite? Oh, that's when you're crammed into the back seat of a very small car, feeling a little carsick, on a narrow, windee road, with six very large geologists, and somebody yells "There's a garnet peridotite!," and they screech to a stop and climb out over you and start hammering on the rocks. That's a garnet peridotite.

Many volcanoes occur in arcuate groups over so-called subduction zones, zones where oceanic lithosphere is drawn down into the Earth's mantle. Basalt is common here, but more siliceous rocks, called andesite and rhyolite, are also prominent. Their lavas tend to be less fluid than basaltic lavas. They are often fragmental. Their volcanoes tend to be steep-sided cones, called *stratovolcanoes*, one well-known example is Mt. Fujiyama.

Comparing Peridotite, Basalt, Andesite, and Rhyolite

Rock Name	*Essential Minerals*	*SiO_2 Content*
Peridotite	Olivine, pyroxene	40 percent
Basalt	Pyroxene, plagioclase feldspar	50 percent
Andesite	Plagioclase feldspar	60 percent
Rhyolite	Alkali feldspar, quartz	70 percent

Why are magmas over subduction zones different from magmas at mid-ocean ridges? There are three possibilities, which may all be true. First, the magmas are created by partial fusion of the down-going oceanic crust, which is basaltic. (Mid-ocean ridge basalts are produced by partial melting of peridotite.) Second, the magmas may contain materials melted from the overlying continental or oceanic crust. And third, the magmas typically contain a few percent water. Melting in the presence of water tends to increase the silica content of the melt.

Subduction zone volcanic eruptions can be extremely explosive and dangerous (consider Mt. St. Helens in 1987); and estimates of total energy released in some historic eruptions are sobering. The largest known, the eruption of the Philippines volcano Tamboro in 1815, is estimated to have released energy equal to the explosion of 20 billion tons of TNT and to have ejected more than 150 cubic kilometers of material. Tamboro was truly extraordinary, but over the past two centuries there have been more than a dozen eruptions that exceeded the power of 2 million tons of TNT. The agents of this explosive energy are expanding gases, principally carbon dioxide, water, or both. Water is especially important in the highly siliceous magmas of subduction zone volcanism.

A silicate melt can retain a lot of water (up to 10 percent) only if it is confined under pressure. If the pressure is released suddenly, either through eruption or reduction of the overburden, superheated steam explodes from the melt and, simultaneously, the melt freezes and breaks into tiny fragments. The deposits resulting from these explosive eruptions are heaps of particulate matter, called *ash* and *tuff*. Usually such deposits remain loose like unconsolidated sediments, but sometimes sufficient heat is retained (or generated by local oxidation) to actually weld the particles together, forming a very tough, hard rock called *welded tuff*. Many welded tuffs have the composition of rhyolite.

Although the name *rhyolite* was created from the Greek word meaning stream or flow, most rhyolite formations are not frozen lava but deposits of solid particles that have been welded together. This occurs because molten rhyolite is almost too viscous to flow. Because rhyolite contains half again as much silica (SiO_2) as basalt, and because rhyolitic magmas normally exist at temperatures 200 or 300 degrees below the temperatures of basaltic magmas (e.g., 850°C versus 1100°C), the rhyolitic magmas are a thousand times more viscous than the basaltic. You could walk on a lake of molten rhyolite, if you could keep your feet cool. For this reason rhyolitic lava flows are generally small and stop close to, or within, the volcano, rarely flowing farther than a kilometer or two. Longer flows, such as the five-kilometer flow in the Medicine Lake Highlands of California, might be evidence that exceedingly viscous melts may become superheated, that is, significantly hotter than the temperature at which they begin to crystallize. Less viscous melts squirt or trickle away from the melting zone before absorbing so much heat.

More common than rhyolite lava flows are stiff spines or domes poking out of volcanoes or glowing avalanches of hot ash called *nuées ardentes*. Nuées ardentes are terrifying, sweeping silently down volcanic slopes as rapidly as 100 miles per hour and searing, choking, or incinerating every creature they encounter. The most famous catastrophe of this sort was the complete devastation of St. Pierre on the Caribbean island of Martinique by a nuées ardente from Mt. Pelée in 1902. This event, on the morning of May 8, killed 28,000 people in less than two minutes. One evening two months later, from a ship lying off of St. Pierre, two geologists from the Royal Society of London watched a similar eruption:

> As the darkness deepened, a dull-red reflection was seen in the trade-wind cloud which covered the mountain summit. This became brighter and brighter and soon we saw red-hot stones projected from the crater, bowling down the mountain slopes and giving off glowing sparks. Suddenly the cloud was brightly illuminated . . . In an incredibly short space of time a red-hot avalanche swept down to the sea. . . . It was dull red with a billowy surface, reminding one of a snow avalanche. In it there were large stones, which stood out as streaks of bright red, tumbling down and emitting showers of sparks. In a few minutes it was over.[40]

Horrific as nuées ardentes are, they must be trifles compared to the eruptions that long ago spread rhyolitic deposits across tens of thousands to hundreds of thousands of square kilometers of New Zealand, western North America, and other continental areas. Such eruptions have not occurred in the brief span of recorded history, and how they came to be so widespread, given the magma's viscous nature, was a mystery until 1919 when Clarence Fenner of the Carnegie Institution of Washington studied a 7-year-old ash deposit in Alaska's Valley of Ten Thousand Smokes. Fenner visualized the phenomenon as a "sand flow" of incandescent particles that expelled gas as they were carried along in a fluidized (gas-supported) mass. A few years later, New Zealander Philip Marshall proposed the name *ignimbrite* for such rocks. Ignimbrite flows are thought to originate from large, near-surface bodies of wet, silica-rich magma. They are presumed to flow rapidly, as boiling foam or fluidized, incandescent sand, from fiery fissures, out across the land for miles and miles. Such apocalyptic phenomena have not been witnessed in human memory, but they will occur again someday.

There are also major, mysterious volcanoes that occur within plates, far from mid-ocean ridges or subduction zones. We do not have a generally accepted explanation for these volcanoes, but a popular one is the *Mantle Plume Hypothesis*, which suggests that the mantle contains plumes of hot material that function as focused heat sources for melting in the asthenosphere (the soft mantle layer beneath the lithosphere); as the lithosphere moves across such "hot spots" a chain of volcanoes such as the Hawaiian-Emperor chain may develop. One test for this hypothesis is that the volcanic activity should be progressive rather than simultaneous, from one end of the chain to the other. The age of volcanism does increase as one moves west from Hawaii, and a baby volcano is now forming on the sea floor southeast of Hawaii.

Plumes may also be responsible for tremendous volumes of basaltic magma that occasionally pour out upon continental crust. Erupting through widespread swarms of parallel fissures, these easy-flowing lavas, called *continental flood basalts*, inundate extensive regions. Individual flows may cover 1,000 square kilometers, and the total extent of the flood lavas may exceed a million square kilometers. Superimposed, successive flows, each 1 to 50 meters thick, may build stacks as thick as 10 kilometers.

Like mid-ocean ridge basalts, continental flood basalts erupt where Earth's lithosphere is being pulled apart (rifted). Unlike the mid-ocean ridge eruptions, which are normally confined to a single, relatively narrow rift valley, broad swarms of parallel fissures occur over continental regions that are tens to hundreds of kilometers wide. Examples of flood basalts associated with major continental rifting events are the basalts of the Lake Superior Province of North America, the Brito-Arctic Province of the North Atlantic, the Paraná Basin of Brazil, the Karoo Province of South Africa, and the Deccan Traps of India.

Continental flood basalt episodes have poured hundreds of thousands of cubic kilometers of lava onto Earth's surface over periods of tens of millions of years. Their production rates of 10 to 30 million cubic meters per year are comparable to the productivity of large oceanic volcanoes, such as those of the Hawaiian Islands. Impressive as they are, we should bear in mind that *all* of the intraplate volcanism (volcanism far from the plate boundaries), central volcanoes, and continental flood basalt eruptions added together are but a small fraction, perhaps only 10 percent, of Earth's magma production.[41] The principal magmatic activity occurs along the mid-ocean ridges, and appears to have done so throughout Earth's history. In fact, the geologic record suggests that continental flood basalts may have been less common in the distant past. The oldest known continental flood basalts erupted only 1,200 million years ago, when Earth was already more than 3,300 million years old, and at least five episodes have occurred in the last 250 million years.

Great outpourings of lava, on continents or in the ocean basins, can affect Earth's climate to the point of discomfiting or even extinguishing some species. Eruptions above sea level affect climate by injecting massive quantities of volcanic carbon dioxide and sulfur dioxide into the atmosphere. Below-sea-level eruptions may affect climate by raising the water temperature. Evidence for the latter comes from the southeastern Pacific, where data from 1964 to the present show strong correlations between El Niño episodes and seismic activity along the East Pacific Rise.[42] The East Pacific Rise is the most active mid-ocean ridge on Earth, and earthquakes along it are apparently related to igneous activity.

Four

From Mud to Metamorphism

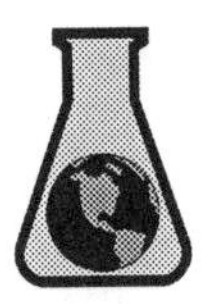

The insight that led James Hutton to his Theory of the Earth—his grand concepts of the rock cycle and geologic time—stemmed from watching the erosion of his Berwickshire farmlands and wondering about the fate of the water-borne mud. Of course, he realized, it would be redeposited in the sea, but what then? To leap from seabed mud to sedimentary rocks, metamorphism, mountain raising, and the rock cycle was the extraordinary intellectual achievement for which he is rightly regarded as the founder of modern geology.

Rocks are formed beneath the surface of the planet, under conditions that are generally less oxidizing and hotter, under greater pressure, and closer to Earth's center of gravity than conditions at the surface. When the rocks are brought to the surface, by volcanism, tectonic processes, and erosion, they undergo changes that we call *weathering*, which can be caused by physical forces, heat, chemical agents, and living organisms. We speak of mechanical and chemical weathering, but it is important to realize that these processes work together.

Fragmentation increases the surface area of a rock and thereby exposes more atoms at the surface, causing them to react, dissolve, or evaporate. When gravel, which contains pebbles 2 centimeters in diameter, is ground to coarse sand particles, which are 2 millimeters in diameter, the surface area of the entire mass increases by an order of magnitude. If the sand is ground down to silt, the surface area is a thousand times greater, and if the particles are reduced to clay size,* the surface area becomes vast indeed—a gram of dry clay, one-third of a teaspoonful, can have several hundred square meters of surface area. Mechanical weathering can reduce particles to clay size by grinding rocks beneath glaciers or along fault surfaces, but this is unusual. Most clay-sized particles are created by chemical weathering.

Water is the most important agent of weathering on Earth, both chemical and mechanical. Water carries the sand and other abrasives that grind rocks down. In temperate regions it shatters rocks because it expands upon freezing; with each freeze, ice pries cracks further open, and with each thaw, water trickles further into the cracks. Because the expansive force is great—approximately 2,000 pounds per square inch—the freeze–thaw cycle eventually breaks all rocks, no matter how massive and strong they may be.

Other agents of mechanical weathering are gravity, organisms, wave action, and wind. Gravity provides force to break rocks apart. Myriad biological agents, from bacteria to rampaging elephants, also cause fragmentation, as do waves, currents, and wind. Thermal cycling, repeatedly hot in daytime and cold at night, causes mechanical weathering if the rocks have a film of water, although without water, thermal cycling does little or no damage. On the Moon, which is bone dry, rocks are subject to large temperature cycles every day (107°C to –153°C), yet many are still intact.

Water's effectiveness as an agent of chemical weathering is due to its polar nature and its ability to form hydrogen bonds.* The polarity results from the placement of both hydrogen atoms on the same side of the water molecule which attract oxygen atoms in adjacent water molecules. Although much weaker than the hydrogen–oxygen bond within the molecule, these intermolecular hydrogen bonds are important. Were it not for them, liquid water would boil at around *minus* 100°C! Water's polar nature makes it a power-

---

*Less than 2 microns or 2×10^{-6} meters.

*A hydrogen bond is an intermolecular, electrostatic attraction between a hydrogen atom in a molecule and a nitrogen, oxygen, or fluorine atom in an adjacent molecule.[43]

ful solvent for ionic compounds. Positive ions, such as sodium (Na^+), are pulled from their crystal by the negative (oxygen) side of the water molecule; negative ions, such as chloride (Cl^-), are pulled away by the positive (hydrogen) side.

Water dissolves carbon dioxide in the atmosphere or in the soil and produces carbonic acid, which is an extremely important agent of chemical weathering:

$$\underset{}{H_2O} + \underset{}{CO_2} \rightarrow \underset{\text{carbonic acid}}{H_2CO_3}$$

Chemical weathering of limestone usually involves carbonic acid:

$$\underset{\substack{\text{calcite}\\\text{(limestone)}}}{CaCO_3} + \underset{\substack{\text{carbonic}\\\text{acid}}}{H_2CO_3} \rightarrow \underset{\substack{\text{dissolved}\\\text{calcium}}}{Ca^{++}} + \underset{\substack{\text{dissolved}\\\text{bicarbonate}}}{2HCO_3^-}$$

Notice that in this process the limestone is completely dissolved. In contrast, the weathering of feldspar by water and carbonic acid leaves a solid residue of clay (kaolin) and quartz:

$$\underset{\text{feldspar}}{2KAlSi_3O_8} + \underset{\text{carbonic acid}}{2H_2CO_3} + H_2O \rightarrow$$

$$\underset{\text{kaolin (clay)}}{Al_2Si_2O_5(OH)_4} + \underset{\text{quartz}}{4SiO_2} + \underset{\substack{\text{dissolved}\\\text{potassium}}}{2K^+} + \underset{\substack{\text{dissolved}\\\text{bicarbonate}}}{2HCO_3^-}$$

The residual kaolin and quartz, which have less than half of the volume of the original feldspar, makes a soil called laterite. Continued leaching of this laterite by groundwater will eventually remove the quartz and transform the kaolin into gibbsite ($Al(OH)_3$), which is the commercial source of aluminum. The soft rock composed of gibbsite is called bauxite.

Weathering of iron-bearing rocks often involves oxidation (i.e., rusting). The iron changes from ferrous iron (Fe^{2+}), whose compounds are green or gray, to ferric iron (Fe^{3+}), whose compounds are red or yellow. The "painted deserts" of Earth and Mars are due to ferric iron.

Progressive weathering etches away the less resistant minerals, leaving the more resistant ones standing out in relief. The term for this process is *differential weathering,* which you can see in many limestones, where sandy patches stand out as resistant bumps. On a larger scale, you can see that many topographic features are strongly influenced, if not actually caused, by differential weathering. From the ridges and valleys of Appalachia to Shiprock in New Mexico, differential weathering shapes the landscape.

Mechanical and chemical weathering transforms rocks into more stable chemical and physical forms, and erosion transports the newly formed materials towards the Earth's center of gravity. This process, which smoothes the Earth's surface, is powered by the planet's two great heat sources. The Sun drives oceanic and atmospheric circulation, and drives the hydrologic cycle that raises water from the oceans up onto the mountains. The Earth's internal heat raises some portions of the Earth's surface above others and creates the mountains. If the Earth's internal heat were exhausted, mountain building and other uplifting processes would cease. Erosion would continue for a time, but eventually the land would be worn down to a uniform level, a few meters below sea level, at the base of oceanic waves. Chemical weathering would continue but very slowly. The only mechanical weathering would be due to burrowing worms and other active organisms. Except for occasional submarine landslides, erosion would cease. The continents would become broad, wavecut platforms surrounded by large ocean basins.

The products of mechanical and chemical weathering—boulders, sand, clay, and ions in solution—are closer to chemical stability than they were before weathering, but they still possess gravitational potential energy that they can release by moving closer to the Earth's center of gravity. So, down the mountains they come, mostly helped along by flowing water, sand and gravel (the bedload) bouncing along stream floors, clay and sand suspended in turbulent currents, and the dissolved substances slipping by, largely unnoticed, in solution. The detrital and dissolved substances vary from place to place, according to the nature of the rocks and the climate.

Major Dissolved Substances (mg/kg)[44]

Substance	*River Water*	*Sea Water*
Sodium	6.9	10,760
Magnesium	3.9	1,294
Calcium	15.0	412
Potassium	2.1	399
Chloride	8.1	19,350
Sulfate	10.6	2,712
Bicarbonate	55.9	145

The quantity of sediment discharging into the oceans also is different from one place to another. It depends upon the size of the drainage basin, the amount of rainfall in the source areas, and the relief. Seventy percent of the world's sediment discharge occurs in southeast Asia. There is little discharge into the Arctic and North Atlantic Oceans, the former because of aridity and low relief, the latter because of barriers erected by people in western Europe and eastern North America.

Water Discharge Compared to Sediment Discharge[45]

River	*Water (m^3/sec)*	*River*	*Sediment (10^6 t/yr)*
Amazon	113,330	Ganges/ Brahmaputra	1,670
La Plata	79,300	Huangho	1,080
Congo	39,600	Amazon	900
Yangtze	21,800	Yangtze	478
Brahmaputra	19,800	Irrawaddy	285
Ganges	18,700	Magdalena	220
Mississippi	17,500	Mississippi	210

In the course of their trek towards the center of gravity, sediments and dissolved substances may be delayed. This may be only a brief delay in a backwater, or it may be a longer delay in a lake or a final halt in an ocean basin. Sedimentation begins when sediments and dissolved substances stop moving.

The dichotomy between fragmental and precipitated rocks, and between mechanical and chemical weathering, persists in the two kinds of sedimentation: *clastic* and *chemical*. Clastic sediments are composed of broken bits and pieces of rocks,* which on Earth are transported by air, water, or ice and deposited from air or water. Deposition occurs when the air, water, or ice can no longer transport the clastic material as a suspended load or bed load. The capacity of the transporting medium increases with volume, velocity, density, viscosity, and the amount of turbulence. It is also affected by the nature of the sediment load: A stone sinks faster if it is bigger, if it is more dense, and if the fluid is less viscous.** There is an inverse relationship between the transporting medium's carrying capacity and its ability to sort particles according to size and density. Air is very effective at sorting grains, so æolian (wind borne) deposits tend to contain grains of similar size and density. On the other hand, sediments carried by ice are poorly sorted mixtures of clay, sand, pebbles, and boulders. Water-borne sediments vary in degree of sorting according to the flow conditions.

There are three principal categories of clastic sediment: mud, sand, and gravel. The resulting sedimentary rocks are, respectively, shale, sandstone, and conglomerate. On Earth the most common of these is shale, which is three times more abundant than the others. Shale is compacted, dewatered mud. In addition to sand and gravel, sandstones and conglomerates contain carbonate or silica and/or some other interstitial material, such as clay, which binds the rock together.

There are many environments where clastic sediments are deposited. The most important are *alluvial*, where deposits are formed in river channels, on alluvial fans where steep, narrow valleys enter broad plains, and on flood plains along the banks of

---

*Compare the word *iconoclast*, which means a breaker of icons.

**This relationship is expressed by Stoke's Law: Settling velocity = $2gR^2(d_1-d_2) \div 9V$, where g is the acceleration due to gravity on Earth, R is the radius of the particle, d_1 is the density of the particle, d_2 is the density of the fluid, and V is the viscosity of the fluid.

rivers; *æolian,* where wind-borne deposits are found in deserts and other arid regions; *glacial,* where deposits are left by glaciers; *deltaic,* where deposits are formed by rivers entering bodies of less active water; *lacustrine,* where deposits are formed in lakes; and *marine,* where deposits are formed in the ocean.

On Earth, chemical sediments are materials that precipitate from water. In order to appear prominently in chemical sediments, the material must be moderately soluble in water. Because of its polar molecule, water is an effective solvent of atoms that are readily ionized. The most easily ionized elements are those that have either a small number of valence electrons, which they lose, or a large number of valence electrons, so that the addition of one or two more achieves a stable octet. Common cations are sodium (Na^+), potassium (K^+), magnesium (Mg^{2+}), and calcium (Ca^{2+}). Common anions are chloride (Cl^-), sulfate (SO_4^{2-}), bicarbonate (HCO_3^-), and carbonate (CO_3^{2-}). Silica dissolves as a neutral species, silicic acid ($Si(OH)_4$), or as part of complex organic molecules. These materials dominate the chemical sediments on Earth.

Earth's chemical sediments are either *biogenic* or *abiogenic* (i.e., not biogenic). Biogenic sediments are extracted from water by plants or animals whether or not the water is saturated with the relevant chemical species. Just as you and I extract materials from our food to build bones and teeth, many organisms that live in water use metabolic energy to draw desired substances into their bodies to form carbonates, silicates, phosphates, and so on. The most abundant biogenic sediments are two forms of calcium carbonate—calcite ($CaCO_3$, usually with a few percent $MgCO_3$) and aragonite ($CaCO_3$). When the organisms die, their hard parts may become carbonate sediments, and these sediments may become limestones. Ten to fifteen percent of all sedimentary rocks are limestones and similar carbonate rocks. Another, less abundant, biogenic sediment is amorphous silica (SiO_2), which is also extracted by organisms to form hard body parts. The resulting sedimentary rocks are called *cherts*. Phosphates, vegetable matter, and hydrocarbons are also biogenic sedimentary materials.

Abiogenic chemical sediments form only when the water is saturated, or supersaturated, with the relevant chemical species. However, although saturation is a necessary condition for precipitation, it may not be sufficient. Kinetic factors, such as the difficulty of nucleation, may inhibit precipitation. A striking example of this is calcium carbonate in seawater.

The surface waters of the oceans are saturated with respect to calcite. In fact, they contain more than twice the concentrations of calcium and carbonate ions needed for saturation. This situation is favorable for extraction of calcium carbonate from surficial seawaters by organisms, yet abiogenic calcite precipitation is rare. The reasons appear to be energetic barriers to nucleation and interference by other dissolved substances, particularly magnesium.[46]

The only prominent abiogenic carbonate precipitates in the ocean are tiny* balls of aragonite or, less commonly, calcite, called *oöids,* a word pronounced "ooh-id" that comes from the Greek word for egg and has the delightful attribute of resembling the object it stands for. Oöids form by direct precipitation of concentric layers of calcium carbonate around minute fragments of preexisting clastic grains. Sedimentary rocks composed of oöids are called *oölites.* Carbonate oöids are the most common variety; oöids are also composed of hematite (Fe_2O_3) and other, more complicated clay-like minerals. Some of these are primary precipitates, and others have replaced pre-existing carbonate oöids.

Warm temperatures, high salinity, and vigorous current or wave action favor abiogenic carbonate precipitation. All of these conditions diminish carbonate solubility in water by decreasing the amount of dissolved carbon dioxide.

Another example of abiogenic carbonate precipitation due to decreased carbon dioxide concentration, which is much more widely known than oölitic limestone, is calcite deposition in caverns. When groundwater saturated with carbon dioxide and calcium carbonate enters an open cavern, it loses carbon dioxide to the air. Calcite precipitates immediately, forming stalactites, stalagmites, and other "speleothems." Here, as in the ocean, equilibrium between dissolved carbonate and dissolved carbon dioxide is regulated by the *Law of Mass Action.*

The Law of Mass Action was proposed by two Norwegian chemists, Cato Maximilian Guldberg and Peter Waage, in 1865. They found by experiment that, at constant temperature, the rate of

*Less than 2 mm, usually 0.2 to 0.5 mm, in diameter.

a chemical reaction in a solution is directly proportional to the mathematical product of the concentrations of the reacting substances.* Chemical equilibrium may be thought of as two reactions, involving the same chemical species, proceeding in opposite directions at equal rates. The Law of Mass Action says that the concentrations of these dissolved chemical species are related by a number, called the *equilibrium constant.* What happens if one or more of the concentrations changes and the equilibrium is perturbed? The system will react so as to counteract the perturbation and restore the equilibrium.

In Earth's natural waters, the solubility of calcium carbonate is related to the presence of carbonic acid, which dissociates into hydrogen ions and bicarbonate ions:

$$H_2CO_3 \rightarrow H^+ + HCO_3^-$$

---

*If $A + B$ react in a solution to form $C + D$, the rate of the reaction is directly proportional to the concentrations of the reactants A and B:

$$\mathrm{A + B \rightarrow C + D}$$
$$V_1 = K_1 \times [A] \times [B]$$

where K_1 is a numerical constant and $[A]$ and $[B]$ are the concentrations of A and B in the solution. Similarly, if the opposite reaction occurs, so that $C + D$ react to form $A + B$, the rate of that reaction is directly proportional to the concentrations of those reactants:

$$V_2 = K_2 \times [C] \times [D]$$

where K_2 is a different numerical constant.

If A, B, C and D are at equilibrium in the solution, the rates of these opposing reactions must be equal. At equilibrium:

$$\mathrm{A + B = C + D}$$
$$V_1 = V_2$$
$$K_1 \times [A] \times [B] = K_2 \times [C] \times [D]$$

Rearranging terms gives the following relationship:

$$K_1 \div K_2 = [C] \times [D] \div [A] \times [B]$$

The ratio K_1/K_2 is the equilibrium constant.

The amount of calcium carbonate dissolved in the solution is related directly to the concentration of hydrogen ions:

$$CaCO_3 + H^+ = Ca^{2+} + HCO_3^-$$

If carbon dioxide is removed from solution, the concentration of carbonic acid will also decrease because of the equilibrium between water, carbon dioxide, and carbonic acid:

$$H_2O + CO_2 = H_2CO_3$$

This will decrease the concentration of hydrogen ions, and calcium carbonate will precipitate.

What if the concentration of carbon dioxide increases? The concentrations of carbonic acid and hydrogen ions increase and the water can dissolve more calcium carbonate. This happens with increasing depth in the ocean as the water becomes colder and therefore able to dissolve more carbon dioxide. In ocean water deeper than about five kilometers (deeper in tropical than in polar oceans), the rate of carbonate dissolution exceeds the rate of carbonate precipitation (biogenic or abiogenic). Below this depth, which is known as the *carbonate compensation depth,* or CCD, carbonate minerals are rare or absent from sediments,* which is why deposits of siliceous ooze and clay dominate abyssal (deep ocean bottom) marine sediments.

The siliceous oozes, which form sedimentary rock called chert, are biogenic. The ocean is not saturated with silica at any depth, but microscopic creatures called *radiolarians* and *diatoms* extract silica from the water in order to build their skeletons. When they die, their siliceous skeletons rain down upon the seafloor.

A special kind of chemical sediment is produced when a sea or lake dries up, which has happened often in Earth's history. A spectacular instance occurred between 5 and 15 million years ago when the Mediterranean Sea was cut off from the Atlantic Ocean. Rainfall and rivers did not supply enough water to the Mediterranean to compensate for evaporation, so, within a few thousand years, the sea dried up. When seas evaporate various abiogenic chemical sediments

*There are actually two compensation depths. For aragonite, the CCD lies at two kilometers, for calcite it lies at five kilometers.

precipitate from the saltwater. You may be surprised to learn that in spite of the high concentration of sodium chloride in seawater, the first substance to precipitate is not salt but calcium carbonate. The sequence of minerals precipitating from evaporating seawater is calcium carbonate (calcite), calcium sulfate (gypsum), and then salt (halite).

Almost all sedimentary rocks are mixtures of clastic materials (those composed of bits and pieces of pre-existing rocks) and chemical precipitates. Sandstones, for example, are commonly held together by a carbonate or silica cement, and almost all limestones are composed of biogenic carbonate detritus, ranging in size from submicroscopic aragonite needles, derived from calcareous (calcium carbonate bearing) algae, to mountainous heaps of reef-forming organisms. Many limestones also contain grains of quartz sand, clay, or other noncarbonate minerals. Clastic materials and chemical precipitates have different tales to tell. The clastic materials tell about the provenance—the surroundings—of the sedimentary environment. Were there volcanoes or mountains of metamorphic rock close by? Was the sediment deposited close to shore or far out in the open ocean? The clastic grains will tell. The chemical precipitates tell about the depositional environment and the history of the sediment after deposition.

As soon as sediment accumulates, it begins to change by compaction, recrystallization, dissolution, replacement, and precipitation, in response to increasing pressure and temperature and changing pore fluids as the sediment is buried deeper and deeper. This process is called *diagenesis*.

Compaction is accomplished by rearranging, bending, and dissolving clastic grains. During the first million years, as the depth of sediment burial increases to a hundred meters or so, much of the interstitial water is squeezed out. For mud this may amount to as much as 60 percent of the deposit's original volume. With increased depth, temperature, and pressure, compaction is accompanied and enhanced by dissolution of clastic grains. Sometimes zones of dissolution are marked by thin layers of less soluble mineral grains that were left behind. When viewed on a plane surface, these thin layers resemble lines drawn by a trembling hand and are therefore called *stylolite* seams.*

Initially the interstitial waters reflect the chemical conditions of the depositional environment. If the environment is a wave-washed carbonate platform, such as one might visit in the Bahamas, the interstitial waters will resemble well-oxygenated seawater; but if the environment is a fetid swamp, the waters will contain little dissolved oxygen and may contain hydrogen sulfide instead. Soon after sediment burial, however, the interstitial waters begin to change. Free oxygen is likely to be consumed in oxidation of organic matter among the sediment. As burial continues, more clastic grains are dissolved and the pore fluids acquire higher concentrations of dissolved silica, calcium carbonate, iron, and other substances that they are likely to reprecipitate as interstitial cement and as overgrowths on other clastic grains. Eventually, some pore waters acquire such high concentrations of dissolved alkali metals, aluminum, and silicon that they precipitate clay minerals and even alkali feldspars.

Minerals in the sediment may alter their crystalline structures and chemical compositions after deposition. This process, called *recrystallization*, is especially common among the clay and carbonate minerals, and, among the latter, a most interesting process is the transformation of calcium carbonate into dolomite.

Dolomite is a highly ordered, calcium magnesium carbonate ($CaMg(CO_3)_2$), which gives its name to carbonate sedimentary rocks of the same composition. Dolomite the rock appears with increasing frequency as one looks farther back in the past. Limestones dominate the recent past, but dolomites are more abundant than limestones among rocks more than 550 million years old. Most—perhaps all—dolomite forms by replacing preexisiting calcium carbonate. In some instances the replacement process is almost instantaneous, occurring before the original carbonate is buried. The replacement process requires water with a high ratio of magnesium ions to calcium ions and low sulfate ion content, and

---

*Stylolites are common features of many ornamental marbles. The slabs of pink, gray or white marble, which were often used as partitions in public restrooms before the introduction of stainless steel and plastics, are usually decorated with stylolites.

there seem to be several ways to create such water. Seawater itself may be appropriate, especially if bacteria remove the dissolved sulfate; dolomite formation has been observed in carbonate oozes on shallow, submarine slopes of the Bahamas. Mixtures of fresh and sea water may also be suitable, as may pore fluids expelled from deeply buried shales. Whatever the source of the dolomite-forming fluid, however, substantial volumes of water must pass through the carbonate sediment or limestone in order to supply the necessary magnesium and remove the excess calcium.

Compaction, recrystallization, dissolution, replacement, and precipitation continue as the sediments and sedimentary rocks are buried, and after a few million years the rocks may be buried beneath several kilometers of younger sediments. Temperatures may exceed 200°C, and pressures may exceed 1,000 bars. Clays and other minerals of Earth's surface begin to be transformed into different compounds, which are stable under the conditions of higher temperature and pressure. At this point the rocks have entered the realm of metamorphism.

James Hutton related unconsolidated sediments on the seafloor to consolidated sedimentary rocks in which one might still recognize objects such as fossils, and he related sedimentary rocks to rocks such as marble, the metamorphic equivalent of limestone, in which all sedimentary objects may have been obliterated. In making these connections he described diagenesis, or the processes that create sedimentary rocks from loose sediment, and he implied the process of metamorphism, which creates marble from limestone, slate from shale, schist from slate, and so forth. No one before had imagined that a rock might be transformed; it was one of the few unprecedented ideas in science. Although Hutton did not clearly distinguish igneous and metamorphic processes as, respectively, transformations with and without fusion, he did refer to the effects of heat and fusion separately, and he emphasized the role of *change*:

> If, in examining our land, we shall find a mass of matter which had been evidently formed originally in the ordi-

> nary matter of stratification,* but which is now extremely distorted in its structure, and displaced in its position,—which is also extremely consolidated in its mass, and variously changed in its composition,—which therefore has the marks of its original or marine composition extremely obliterated, and many subsequent veins of melted mineral matter interjected; we should then have reason to suppose that here were masses of matter which, though not different in their origin from those that are gradually deposited at the bottom of the ocean, have been more acted upon by subterranean heat and the expanding power,* that is to say, have been changed in a greater degree by the operations of the mineral region.[47]

Like many other persons of genius, Hutton did not find his niche immediately. He was attracted to chemistry at the University of Edinburgh, but was directed into apprenticeship to a lawyer. That didn't work, so he took up medicine; however, when his medical studies were completed Edinburgh had no need for yet another physician, so he turned to farming in nearby Berwickshire. A short way from his farm, on the coast, lived the young baronet of Dunglass, Sir James Hall (1761–1832).

Sir James was not the first experimentalist in geology, but because of his relationship to James Hutton he is acclaimed as the founder of experimental geology. Actually, the high temperature experimental study of rocks began with metallurgy (ca. 7,000 B.C.) and experienced a major acceleration with the creation of porcelain. Made from pure kaolin at temperatures around 1,250°C, porcelain was invented in China before the third-century A.D. but not duplicated in Europe until the eighteenth century. Beginning then, the lessons of its manufacture, that rock-like crystalline materials could be created by controlled cooling of silicate melts, was sporadically applied to geological questions by various individuals, but their influence was negligible. The reason seems to have been an

---

*Sedimentary rock layers are called strata, so the term "stratification" refers to the original sedimentary layering. The "expanding power" refers to the power of Earth's internal heat to uplift and deform rocks.

assumption of irrelevance. Hutton himself had little or no use for experimentation and decried those "superficial reasoning men, who . . . judge of the great operations of the mineral kingdom, from having kindled a fire, and looked into the bottom of a little crucible."[48]

Nevertheless, Hutton needed experimental support rather desperately. In *Theory of the Earth* he had suggested that the combination of heat and confining pressure on lime mud would produce limestone and marble, yet, to the contrary, Professor Joseph Black at Edinburgh University had demonstrated that intense heat would decompose calcium carbonate into calcium oxide and carbon dioxide. Hall urged Hutton to test the effects of pressure on this reaction, but Hutton refused, according to Hall, "on account of the immensity of the natural agents whose operations he supposed to lie far beyond the reach of our initiation; and he seemed to imagine that any such attempt must undoubtedly fail, and thus throw discredit on opinions already sufficiently established, as he conceived, on other principles."[49] Hutton died in 1797, and Sir James thereupon undertook experimental studies of the effects of high pressure and temperature on limestone. He began by heating crushed limestone in plugged gun barrels, several of which exploded violently.

The potential for explosions has always given high-pressure research a certain *panache*. Stories abound of roof-lifting blasts, and even the most able, experienced, and cautious experimenter can be surprised. The experience of young Joe Chernosky is a case in point:

Thirty years ago, I was fortunate to have a post-doctoral appointment at the Geophysical Laboratory of the Carnegie Institution of Washington, where brilliant scientists such as Norman Bowen had established and were continuing a great tradition in experimental studies. I was sitting in my office one quiet evening when a tremendous boom shook the windows. A noise so loud could not be ignored, and I ran into the corridor to find Joe Chernosky, a predoctoral student from MIT, wide-eyed, deathly pale, and shaking with fright. Poor Joe had been building the gas pressure in Hatten Yoder's apparatus to 10 kilobars when the gland nut, a massive piece of threaded steel the size of a softball, succumbed to metal fatigue, sheared its threads, and was blown through the laboratory wall, landing 20 feet away on Yoder's desk. Yoder, fortunately, was at home that evening. Chernosky, who, sensibly, had been behind a steel safety shield, was temporarily deafened and terrified by the blast, but he resumed his experiments the following day—after Yoder's wall had been patched and reinforced.

Dunglass Castle, the first high-pressure laboratory, itself has an explosive history: During the reign of Charles I, on August 30, 1640, to be precise, the entire castle blew up, killing the rebellious Earl of Haddington along with a dozen other Covenanters who were sworn to oppose popery. Several dozen servants also were killed, and many others "sore hurt." The alledged villain was one Edward Paris, an English page, who may have intentionally touched off the substantial store of gunpower in a vault beneath the castle, but the page could not be interrogated afterwards for "no part of him was ever found but ane arm, holding ane iron spoon in his hand."[50] Such was the heritage of high-pressure research.

Eventually, in safer and more sophisticated experiments, Sir James packed porcelain tubes with fine chalk ($CaCO_3$), sealed the tubes with a gas-tight iron bungs in precisely bored blocks of iron, and heated the assemblies in a glass-making furnace. The results provided a substantial boost for Hutton's theory, for when the bombs were opened the chalk was found to have been metamorphosed into coarsely crystalline, synthetic marble. Hall's synthesis of marble was the first in a long succession of experimental studies that have now defined, rather precisely, the conditions of formation of virtually all known metamorphic rocks.

Prograde metamorphism is change associated with increases in pressure and/or temperature. Heating transforms minerals into other compounds that are more stable at higher temperatures. A common type of reaction breaks down compounds that contain volatile components.

$$\underset{\text{mica}}{KAl_3Si_3O_{10}(OH)_2} \rightarrow \underset{\text{feldspar}}{KAlSi_3O_8} + \underset{\text{corundum}}{Al_2O_3} + \underset{\text{water}}{H_2O\uparrow}$$

The water produced in the reaction above drifts away to regions of lower temperature and pressure, leaving behind a hotter, drier metamorphosed rock. Notice how this reaction is reminiscent of weathering in reverse.

Another general effect of thermal metamorphism is to produce larger individual crystals. Thus, the progressive metamorphism of shale, which is composed of submicroscopic clay crystals, proceeds to slate, wherein the crystals are barely microscopic; to phyllite, where mica crystals, discernible with a hand lens, replace the clay minerals; and then to schist, where individual crystals are apparent to the unaided eye. The next step, from schist to gneiss, may also involve crystal enlargement, but the principal change is to expel water and thus convert hydrous (water-bearing) minerals, such as mica, to anhydrous (water-free) minerals, such as alkali feldspar.

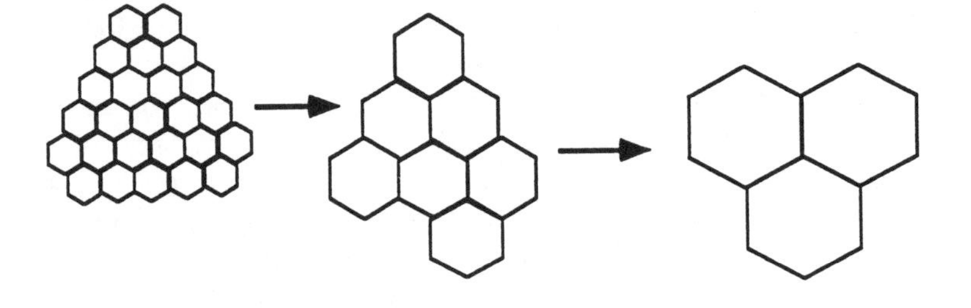

If possible, any polycrystalline aggregate will transform into an assemblage of fewer and larger grains. This tendency, known to metallurgists as *Ostwald ripening*, is driven by the greater stability (reduced free energy) of material objects with reduced surface area. Many rocks form too quickly or are too cold for diffusing atoms to organize as large crystals. This is why lavas tend to be fine grained. Metamorphism, like intrusive igneous activity, provides both time and energy for larger crystals to grow.

Compression tends to make matter more compact, more dense. It is of two kinds: hydrostatic and nonhydrostatic. Hydrostatic pressure exerts equal pressure in all directions and is the kind you feel at the bottom of a swimming pool. An increase in hydrostatic pressure causes reduced volume and higher density, but no change of shape. For example, if we take a normal basalt and subject it to a substantial increase in hydrostatic pressure, say from 1 bar to 20 kilobars, the basalt will change into a more dense rock called *eclogite*:

Rock	*Essential Minerals*	*Density*
Basalt	Plagioclase + pyroxene	2.8 g/cm^3
Eclogite	Garnet + pyroxene	3.4 g/cm^3

The associated decrease in volume is about 20 percent. This reaction turns out to be a common one in Earth's upper mantle, because it occurs wherever the oceanic lithosphere descends into the mantle. As a matter of fact, this reaction, and the associated increase in density, may cause the lithospheric slab to sink as deep as it does.

Nonhydrostatic pressure is directed stress—a push from some direction—that leads to change in shape. Given time and sufficient nonhydrostatic stress, even rocks change their shapes. A fascinating aspect of this phenomenon is that we can look at a squashed object, such as a deformed rock, and determine the direction and nature of the force that did the squashing. This has been one of our principal avenues to understanding how mountains are built.

Retrograde metamorphism, which is caused by cooling or decompression, may not proceed as quickly as prograde metamorphism because the rocks may contain less water, and movement through a film of water along grain boundaries and cracks is the fastest way for atoms to move within rocks. Rocks subjected to prograde metamorphism for the first time, such as shales and other sedimentary rocks, are likely to contain a lot of water. This water is driven off by the increasing temperatures of prograde metamorphism, so when temperatures decrease and retrograde metamorphism begins, the rocks no longer contain as much water, and the atoms have a more difficult time moving.

A major exception occurs in the oceanic lithosphere where pervasive retrograde metamorphism is caused by seawater flowing through the cooling pillow lavas and underlying rocks as they are displaced from the mid-ocean ridge. The circulating seawater reenters the ocean through springs on the seafloor. Studies of heat flow and bathymetry (the measurement of water-body depth) suggested the presence of these springs several years before they were actually observed. The data indicated that rocks along the mid-ocean ridges were losing a lot of heat by some means other than simple thermal conduction. Convective heat transfer by water circulation was the only likely alternative, so marine geologists looked along the ridges for hot springs.

Hot springs were first found on a mid-ocean ridge 280 kilometers northeast of the Galapagos Islands in 1977. Their general location was identified by a tiny temperature rise (less than 0.1°C) in the deep ocean water, and by water composition that indicated the presence of hydrothermal waters mixing with the sea. Then photographs of the normally bleak and barren deep ocean floor revealed astonishing fields of white clams as large as dinner plates, so three men in the deep submersible *R.V. Alvin* went down to take a look. Dropping below the green, sunlit surface water, they sank slowly through a mile and a half of total darkness. After an hour and a half they leveled off a few meters above the ocean floor, switched on the search lights, and looked out. The first sign of hot springs was water shimmering above the basaltic pavement, like hot air rising off a summer road. One of the three men, geochemist John Edmond of MIT, described the scene:

> Reefs of mussels and fields of giant clams were bathed in the shimmering water, along with crabs, anemones and large pink fish. . . It soon became apparent that we had come on a hot-spring field. Warm water streamed from every orifice and crack in the sea floor over a circular area about 100 meters in diameter. The temperature of the water was highly variable, but the maximum was about 17 degrees C. The organisms were quite selective. They choked the warmest vents. In some cases mussel reefs actually channeled the flow of water, forming conduits themselves.[51]

The clams and other organisms were members of a previously unknown ecosystem that depends for its energy not on the Sun but on Earth's internal heat, delivered in complete darkness through submarine hot springs. The total amount of water flowing through such springs is very large, in fact, comparable to the Mississippi River. The entire ocean (1.37×10^{21} liters) circulates through them on the worldwide mid-ocean ridge system every 8 to 10 million years. The amount of Earth's internal heat issuing from these springs worldwide is also very large—perhaps as much as 10 percent of the heat that flows to the planet's surface.

The field of springs that Edmond and his colleagues discovered on the Galapagos Ridge turned out to be rather placid and cool compared to the springs that were discovered about a year later by French and American scientists on the East Pacific Rise, another mid-ocean ridge, near the mouth of the Gulf of Mexico. At first the divers saw features and creatures similar to those seen earlier on the Galapagos Ridge:

> Shimmering water rose between the basaltic pillows along the axis of the neovolcanic zone. Large white clams as much as 30 centimeters long nestled between the black pillows; white crabs scampered blindly across the volcanic terrain. Most dramatic of all were the clusters of giant tube worms, some of them as long as three meters. These weird creatures appeared to live in dense colonies surrounding the vents, in water ranging from two to 20 degrees C.

But then *Alvin* dove into a nearby hydrothermal area where the scene was very different and very dangerous, inspiring comparisons with fire hoses, steam locomotives, or Pittsburgh of the 1930s:

> . . . extremely hot fluids, blackened by sulfide precipitates, were blasting upward through chimney-like vents as much as 10 meters tall and 40 centimeters wide. We named the vents "black smokers." The chimneys protruded in clusters from mounds of sulfide precipitates. . . . Our thermometer was calibrated to 32 degrees C; when it was inserted into the first chimney, the readings immediately sailed off the scale. . . . the plastic probe on which [the thermometer] was mounted showed signs of melting! . . . measurements were made on several more dives; they indicated temperatures of at least 350 degrees C.[52]

As seawater flows through the hot basalts and underlying peridotite, it heats up and exchanges chemical elements with the rocks. Sodium and magnesium go from the water into the rocks, which also absorb a lot of the water, and the hot seawater dissolves many elements from the rocks, including calcium, iron, manganese, copper, lead, and zinc. When this metal-laden saline solution erupts into the ocean it reacts immediately with dissolved sulfate ions in the cold seawater, precipitating clouds of crystals that create white or black "smoke" plumes above the vents. The precipitates rain down on the seafloor, as far as 50 meters from the vents, or build chimneys around the venting fluid streams. Both the blankets of metal-rich precipitates and the chimneys may eventually become rich ore deposits. Such deposits are found in Cyprus (named for copper), Newfoundland, and many other places where ancient oceanic crust has been pushed up and exposed on Earth's surface.

What happens to the basalts? They are transformed into metamorphic rocks containing iron- and magnesium-bearing micas, sodium-rich feldspar, and other metamorphic minerals. Often this metamorphism takes place without destroying the large-scale volcanic structures, such as the pillows.

The underlying peridotites, which are the residual solids left after melting, are also metamorphosed by contact with the circulating water. The result is a beautiful dark-green rock called serpentinite, known commonly as soapstone or verde antique marble. Serpentinite is created when olivine reacts with hot water to form the mineral serpentine, which is soft and slippery and therefore easily cut, carved, and polished. Because it is easily worked and beautiful when polished, serpentinite is a popular ornamental stone often used in banks and hotels, where its rich green color provides striking contrast with lighter shades of marble. Imagine—what is now a rich green, polished column, floor, or table top was once, millions of years ago, a white hot mass of mantle peridotite, bleeding molten basalt beneath a mid-ocean ridge.

The distribution of metamorphic rocks on and within a planet reveals some of the planet's inner workings. In a fundamental way the recognition of metamorphosed rocks, as on Earth, indicates that a planet *has* (or had) "inner workings," that it has (or had) sufficient internal energy to shift substantial quantities of heat and rock from one place to another. For such rocks to be recognized, of course, they must be exposed at the surface. Whether or not they are exposed depends upon an exposure mechanism such as tectonic uplift, weathering, and erosion. The absence of regionally metamorphosed rocks on a planet's surface may indicate a lack of internal energy or want of an exposure mechanism.

Some metamorphic rocks will never be exposed, yet we must understand them to know how a planet works. Deep in Earth's mantle, for example, beneath the lithosphere, where every rock is incandescent, huge metamorphic reactions accompany convective motions as hot rocks at high pressure are raised, slowly cooled, and decompressed—and, conversely, as rocks descend deeper into the planet. We have few opportunities to observe the products of these deep intraplanetary retrograde and prograde metamorphisms, but we can simulate them in the laboratory and ponder their importance.

Five

Older Than the Hills

"This earth," wrote James Hutton, "like the body of an animal, is wasted at the same time that it is repaired."[53] All that we see—the soil, sands, loose stones, and bedrock—and all that is hidden from view—the heating, hardening, bending beds of sediment and the buried masses of hot rock and melt—relate to one another in an endless cycle. There was, so far as Hutton could tell, no primary rock, no parent from which all others derived. We know now that there *was* a beginning, although all vestiges of that infant Earth have long since been transformed. Our knowledge of Earth's history came slowly, beginning with the recognition and determination of relative geologic age and consummated by the measurement of absolute geologic age.

Implicit in James Hutton's concept of the rock cycle is the notion of relative age, that rocks precede and succeed one another in time. Hutton did not originate this idea, some of the fundamental concepts were formalized in 1669 by Nicolas Steno, the same physician who also made important contributions to crystallography and anatomy. Walking in the hills of Tuscany, Steno noticed shark's teeth embedded in the rocks and asked himself how an object such as a fossil could be embedded within a stone.

There is no single path to truth in science. Some insights, like those of Steno's on sedimentary processes, derive from questions that ordinary people do not think to ask. Copernicus asked, "What if the Sun were at the center?"; Einstein asked, "What would the world look like if I rode on a beam of light?" Other insights happen when someone notices something most people would ignore. Newton noticed that the spectral image on his wall was oblong, not circular; Jenner noticed that cowpox discourages smallpox. Still other insights happen when someone has a really clever idea, such as Von Laue's diffraction of X rays, and some happen when a new technique enables new observations. There is no single path to truth in science, but there is a common toll: Beyond the burden of endless testing and the presumption of fallibility, all great achievements in science require prodigious effort. After the extraordinary question is posed, the unusual aspect noted, the clever idea imagined, or the unprecedented view observed, a *mighty* effort is required to significantly advance human knowledge. To suppose otherwise is arrogant and naïve. "There is no substitute for hard work," wrote Thomas Edison. "Genius is one percent inspiration and ninety-nine percent perspiration."[54]

Nicolaus Steno asked a simple question: How do shark's teeth come to be embedded in rocks? And in the process of answering it, he developed three principles of stratigraphy: *The Principle of Original Horizontality*, that most sedimentary beds were originally horizontal; *The Principle of Original Continuity*, that adjacent but discontinuous outcrops of the same sedimentary rock may once have been joined by rocks that were eroded away; and *The Principle of Superposition*, that younger sedimentary rocks lie on top of older sedimentary rocks. Steno's principles provided the means to establish sequential ages among sedimentary rocks, but they did not indicate the time intervals involved because rates of sedimentation and erosion vary. And Steno had no special insight on this matter. His life was woven with the warp and woof of his time, mixing the rational processes of observation, experiment, and analysis with the implacable dogmas of the Church. He accepted the Biblical account of history and assumed that all of the features he had observed—sea shells in the mountains and so forth—dated from Noah's flood. In fact, he joined the Roman Catholic Church in 1667, became a bishop in 1675, died in ascetic poverty, and was beatified in 1988.

Steno's principles provided no way to correlate between dissimilar sediments of the same age or to distinguish between similar rocks of different age. That achievement had to await knowledge that developed during the eighteenth century, and the Industrial

Revolution, which had a voracious appetite for coal. In England canals were needed to move unprecedented tonnage of coal from mines to mills. William Smith (1769–1839), a surveyor and drainage engineer employed to lay out the early canal routes, was also interested in geology, and in the course of his work he learned to use fossils as well as rock types to make geological correlations from one place to another. Smith's correlations took the form of geological maps. As the eighteenth century ended he drew the first geological map of England.

A geological map is one of our most concentrated forms of expression. Not only do such maps contain a large body of factual data, but they also represent intricate and detailed interpretations of natural history, and they tend to be works of great beauty. Geological maps show where different kinds of rocks are on Earth's surface, and they correlate rocks that are related in origin but have been separated by erosion. They show structural features, such as the orientation of tilted beds or the location of faults, they show the relationships of rock types, rock ages, and geological structures to topography, vegetation, or cultural features such as villages, highways, mines, or quarries. Geological mapping was the vehicle that brought geology out of the fog-laden swamps of dogma and conjecture and that, more than any other device, provided the means for geologists to make, record, and communicate quantitative measurements and reproducible observations.

Geological maps are usually accompanied by diagrammatic tables known as *stratigraphic columns,* which show the nature and relative ages of the rocks. If the rocks are sedimentary, successively younger ones appear higher in the column. Igneous rocks may indicate their relative ages by stratigraphic position if they are surficial flows or by cross-cutting relationships if they are intrusions; the rock that cuts across is the younger one. In a stratigraphic column younger intrusive rocks are typically shown cutting across the units on which they intrude. Sometimes there are several generations of cross-cutting veins or dikes.

Determining relative ages from field relationships is not always easy or unequivocal because of inadequate exposures or breaks in the geological record. Suppose we climb three hills, in a region where sedimentary rocks occur in simple, horizontal layers, and draw stratigraphic columns for each hill in our notebook.

Hill A	*Hill B*	*Hill C*
Conglomerate Z	Sandstone Y	Conglomerate Z
Sandstone Y	Shale X	Shale X
Shale X	Sandstone W	Limestone V

Wherever possible we determine the equivalence of the beds on the different hills by observing that they contain the same fossils. After climbing up hill A and hill B we can apply Steno's principles and derive the following stratigraphic sequence:

Youngest bed	Conglomerate Z
	Sandstone Y
	Shale X
Oldest bed	Sandstone W

When we start up hill C, however, we encounter a problem. Is limestone V older or younger than sandstone W? Unfortunately, Sandstone W is completely devoid of fossils, so paleontology is no help to us. We realize that we can not answer this question unless we find both limestone V and sandstone W in the same outcrop.

Farther up hill C we encounter another problem. We can't find sandstone Y, and after searching carefully we determine that it is missing. Conglomerate Z is lying directly on top of shale X. We notice also that the top of shale X isn't as even as it was on hills A and B. It looks as if it had channels gouged out of it, and the channels seem to be filled with conglomerate Z. The boundary between shale X and conglomerate Z is called an *unconformity*, indicating an interruption in the sedimentary record. A break in sediment deposition is often accompanied, as in this case, by a period of erosion.

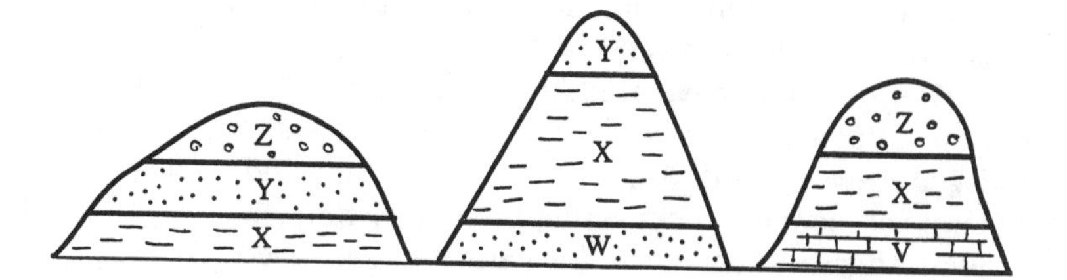

If the beds above and below an unconformity are not parallel, that unconformity is angular, indicating the following sequence of events: First, deposition of sediments; next, deformation (tilting) of the sediments; then, uplift and erosion of the sediments; and, finally, subsidence and renewed deposition. The unconformity surface represents a period of geologic time for which, at least in this place, there is no record. Finding one is like finding a book with some chapters torn out or never written. Some unconformities represent short breaks, but others represent hundreds of millions of years.

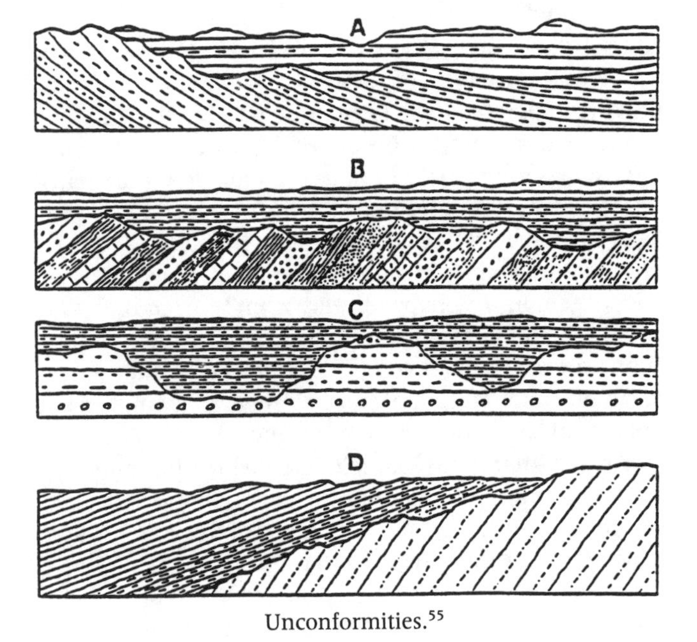

Unconformities.[55]

James Hutton was not the first person to observe and describe an angular unconformity, but he was among the first to comprehend its meaning. His friend and biographer John Playfair (1748–1819) described a day in June of 1788 when he, Hutton, and James Hall, in an adventure reminiscent of Kenneth Grahame's *Wind in the Willows,* explored the coast near Dunglass Castle in a small sailboat. At a place called Siccar Point, they found horizontal beds of the upper Old Red Sandstone lying in angular unconformity upon nearly vertical beds of older shales and sandstones. Playfair, in the role of the ebullient Mole opposite Hutton's Toad and Hall's Water Rat, was quite transported by the experience:

> On us who saw these phenomena for the first time, the impression made will not easily be forgotten. The palpable evidence presented to us, of one of the most extraordinary and important facts in the natural history of the earth, gave a reality and substance to those theoretical speculations, which, however probable, had never till now been directly authenticated by the testimony of the senses. We often said to ourselves, What clearer evidence could we have had of the different formation of these rocks, and of the long interval which separated their formation, had we actually seen them emerging from the bosom of the deep? We felt ourselves necessarily carried back to the time when the schistus [the older shales and sandstones] on which we stood was yet at the bottom of the sea, and when the sandstone before us [the Old Red Sandstone] was only beginning to be deposited, in the shape of sand or mud, from the waters of a superincumbent ocean. An epoch still more remote presented itself, when even the most ancient of these rocks, instead of standing upright in vertical beds, lay in horizontal planes at the bottom of the sea, and was not yet disturbed by that immeasurable force which has burst asunder the solid pavement of the globe. Revolutions still more remote appeared in the distance of this extraordinary perspective. The mind seemed to grow giddy by looking so far into the abyss of time; and while we listened with earnestness and admiration to the philospher who was now unfolding to us the order and series of these wonderful events, we became sensible how much farther reason may sometimes go than imagination can venture to follow.[56]

The unconformity that so moved Playfair and his companions was just one of myriad gaps in the geological record. One who clearly recognized the frequency of such gaps was Charles Darwin, for whom the incompleteness was a serious problem. Models for the evolution of life on Earth are limited by the extent to which sedimentary rocks represent geologic history. Is the sedimentary record more or less complete? Apparently less. Variations in sedimentation rate and the subsequent ravages of erosion result in stratigraphic sections that are likely to be incomplete. Many sedimentary successions have forgotten more than they remember, and this "geological amnesia" must limit the depth to which we can ever comprehend earth history.

During the nineteenth century field geologists established the stratigraphic sequence called the *Geological Time Scale*. Some of the periods were named for regions, such as Devon, where rocks of Devonian age were first studied in detail. Other names were drawn from antiquity: after Cambria, the Roman name for ancient Wales, or after the Ordovices, a Celtic tribe in northern Wales. Still others like the Carboniferous, were named for some characteristic feature, such as the prevalence of coal beds.

Era	***Period***	***Epoch***
Cenozoic	Quaternary	Recent
		Pleistocene
	Tertiary	Pliocene
		Miocene
		Oligocene
		Eocene
		Paleocene
Mesozoic	Cretaceous	
	Jurassic	
	Triassic	
Paleozoic	Permian	
	Carboniferous	
	Devonian	
	Silurian	
	Ordovician	
	Cambrian	
Precambrian		

The eras and periods of the Geological Time Scale were defined by abrupt changes in the fossil record. Many of these changes were caused by mass extinction events that vacated niches and provided opportunities for biotic newcomers. Mass extinction events have been identified at several times in Earth's history, including the ends of the Cambrian, Ordovician, Permian, Triassic, Jurassic, and Cretaceous periods. Some of these resulted from impacts by comets or asteroids. How often such impacts have influenced the destruction and evolution of life on Earth is subject to debate, but there is little doubt that they have been influential.

The Geological Time Scale provided the chapter headings for historical geology and defined the sequence of major historical episodes, but it did not define the lengths of time within or between these episodes. Uniformitarians, including Hutton himself, thought that the age of Earth must be almost infinite. Moderates, such as Charles Lyell (1797–1875) and his friend Charles Darwin, thought that the age was more likely to be hundreds of millions of years. Then, of course, there were the Biblical literalists, taking their lead from seventeenth-century Archbishop James Usher, who reckoned Earth's birth to have occurred during the "entrance of night preceding the 23rd October" in 4,004 B.C.[57]

Science is borne on the progressive presumption that knowing more is possible. But how much more? When science challenges the boundaries of knowledge, an uneasy feeling comes over us. Today we are startled by physicists who suggest that knowing the extent and ultimate fate of the universe is a legitimate and attainable scientific objective. One hundred and fifty years ago Victorians were startled by revelations of geological time and all that that colossal concept implied. The geologists were startled as well, and many stepped back from the edge of knowing. Lyell himself backed away:

> To assume that the evidence of the beginning or end of so vast a scheme lies within the reach of our philosophical inquiries, or even of our speculations, appears to us inconsistent with a just estimate of the relations which subsist between the finite powers of Man and the attributes of an Infinite and Eternal Being.[58]

It had occurred to several people that the age of the Earth might be estimated by calculating how long it had taken the seas to acquire their salt from river waters. In 1715 Sir Edmond Halley (of cometary fame) suggested that the Earth might be dated precisely by determining the salt content of the sea before and after some period of time, but he was forestalled by analytical imprecision. Then, in 1898 Professor John Joly of the University of Dublin performed this feat and calculated an age of 80 to 90 million years. As data accumulated, however, it became obvious that the ocean doesn't work so simply. Rivers are delivering too much material too fast, and the rates differ from element to element.

Time Required for Rivers to Replace Material Dissolved in Seawater[59]

Species	*Replacement Time*
Cl^-	8.7×10^7 yr
Na^+	5.5×10^7 yr
Ca^{2+}	1.0×10^6 yr
SiO_2	2.1×10^4 yr

Residence time,[60] which equals the amount of the substance present in seawater divided by the amount delivered to the sea per year, is more a function of the removal mechanism than of the supply. Sodium and chloride ions, which have the longest residence times (48 and 73 million years, respectively) are mostly removed by evaporation of spray or by the drying up of a sea like the Mediterranean and the burying of the resulting salt beds beneath younger sediments. Retrograde metamorphism of the oceanic crust also removes sodium ions from sea water. Calcium, with a residence time of 850,000 years, is removed by organisms, but may be redissolved in deep water. Silica, with a residence time of only 6,000 years, is removed by organisms as well. Because these mechanisms vary in efficiency according to numerous variable factors (climate, ocean depth, life forms), the salts of the sea cannot reveal the age of the Earth.

Prior to the discovery of radioactivity, physical models indicated that the age of the Earth could not exceed a few tens of millions or, at most, a hundred million years. Hermann von Helmholtz (1821–1894), one of the founders of thermodynamics, thought that the Sun's light was due to gravitational contraction, and on this basis figured that the Sun and Earth could be only 20 to 40 million years old. Lord Kelvin, assuming that Earth had cooled from a white hot liquid and was continually losing heat by radiation into space, could not allow more than 100 million years for Earth to be as warm as it is today, and he was more comfortable with 20 to 40 million years.

Both men were excellent scientists—their arguments were cogent and computationally sound—and they were dead wrong. Their models failed to satisfy the requirements of the geological record. Fortunately, the geological record was appreciated and forcefully represented by equally able scientists, among them a most eloquent advocate, Thomas Crowder Chamberlin (1843–1928) of the University of Chicago. "The fascinating impressiveness of rigorous mathematical analysis," he wrote, "with its atmosphere of precision and elegance, should not blind us to the defects of the premise that condition the whole process."[61] It was an example of scientists learning more from failure than from nonfailure. Chamberlin knew that Kelvin and Helmholtz must be wrong, but he didn't know why. He suggested, with extraordinary prescience, that perhaps there was some other form of energy yet to be discovered.

Discovering and defining that form—nuclear energy—was the greatest scientific triumph of the early twentieth century, and its geological significance was apparent almost immediately. In 1903 Curie and Laborde announced the observation of spontaneous heat generation by radium. The implications for the age of the Earth (and the Sun) produced a flurry of letters to the journal *Nature*: "Radium and Solar Energy," by W. E. Wilson, on July 9; "Radioactivity and the Age of the Sun," by G. H. Darwin (Charles' son), on September 24; and "Radium and the Geological Age of the Earth," by John Joly, on October 1. In 1904 Rutherford wrote, "the time during which the Earth has been at a temperature capable of supporting the presence of animal and vegetable life may be very much longer than the estimate made by Lord Kelvin from other data." He made this announcement in a lecture that Lord Kelvin attended. Rutherford described the scene:

> I came into the room, which was half dark, and presently spotted Lord Kelvin in the audience and realized that I was in for trouble at the last part of my speech dealing with the age of the earth, where my views conflicted with his. To my relief Kelvin fell fast asleep but as I came to the important point, I saw the old bird sit up, open an eye and cock a baleful glance at me. Then a sudden inspiration came and I said Lord Kelvin had limited the age of the earth *provided* no new source was *discovered*. That prophetic utterance referred to what we are now considering tonight, radium! Behold! the old boy beamed upon me.[62]

Kelvin may have beamed, but he didn't budge. In June of 1906, a year and a half before his death, he expressed his final view in a letter:

> I do not think it at all probable that radioactivity can have been practically influential in connection with the primitive heat of the earth, which I believe to have been due to gravitational energy spent in the coalition of smaller bodies. I scarcely think that radioactivity can possibly have had any considerable influence on the subsequent cooling and shrinkage of the earth in its present condition.[63]

His mind had been made up long before, and a mind even as brilliant as Kelvin's tends to set like cement as it ages.

When nuclear decay occurs by beta emission, a neutron in the nucleus decays into a proton and an electron ($N \rightarrow P^+ + e^-$) and the electron is emitted from the nucleus. The charge on the nucleus, the atomic number, increases by 1, but the mass of the nucleus is virtually unchanged because the electron's mass is so tiny compared to that of the proton.*

---

*The mass of an electron is only 1/2000 as great as the mass of a proton or neutron.

Both carbon-14 and rubidium-87 decay by beta emission. Carbon decays to nitrogen-14, and rubidium decays to strontium-87:

$$^{14}C \rightarrow {}^{14}N + e^-$$
$$^{87}Rb \rightarrow {}^{87}Sr + e^-$$

When nuclear decay occurs by alpha emission, the nucleus emits a particle containing two protons and two neutrons. The nuclear mass, the atomic weight, decreases by 4 units., and the number of protons, the atomic number, decreases by 2 units.

Uranium-238, element number 92, decays to thorium-234, element number 90, by alpha emission:

$$^{238}U \rightarrow {}^{234}Th + {}^{4}He$$

The thorium isotope formed is itself unstable and decays. After numerous alpha and beta decays a stable isotope of lead, ^{206}Pb, is formed and the radioactive decay process ceases.

The rates of nuclear decay are not affected by changes of temperature, pressure, chemical environment, or time—the process is inexorable. Different nuclei have different decay rates., which are expressed as *half lives*, or the time required to destroy half the nuclei present. Half lives vary a great deal; rubidium-87's is 47 billion years, and carbon-14's is 5,570 years.

The basis of radiometric age dating is to determine how many half lives have passed since the material acquired its radioactive atoms. If the number of radioactive parent atoms is N_o at the beginning, and N at any subsequent time, the ratio of these numbers decreases exponentially with the passing of half lives. After a certain number of half lives (Y), the ratio $N/N_o = 1/2^Y$.

N/N_o	*Half Lives*
1/1	0
1/2	1
1/4	2
1/8	3
1/16	4
1/32	5
1/64	6

We learn the ratio N/N_0 by determining the proportions of parent and daughter atoms with a mass spectrometer, which is something like a velodrome in which atoms of the element of interest are ionized and released into a curved tube where they race around like cyclists, propelled by a voltage gradient and deflected by a magnetic field. They "hit the wall" (the detector) at various places depending upon their charges and masses.

Physical scientists, particularly Bertram Boltwood (1870–1927) in the United States and Lord Rayleigh in England, recognized the potential for measuring the ages of rocks, but the analytical obstacles were very great. One who devoted much of his professional life to these problems was Arthur Holmes (1890–1965), who began his career in Rayleigh's laboratory. Working with uranium minerals, Holmes was able to present a time scale in 1911 that was closer to the truth than anyone had ever come before.

It is ironic and humorous that, after nearly a century of fighting with physicists to gain more time for the age of the Earth, many geologists were reluctant to accept the physicist's new offering, radiometric ages, because they seemed *too* long! The ages *were* long—longer than almost anyone had imagined—and in the years to come they would grow longer still.

Six

Folding, Faulting, and Floating

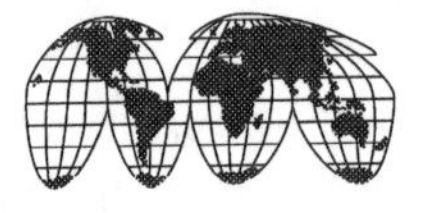

The flagstones on the patio, the slates on the roof, or the boulders in a stream all mislead our intuition about how rocks deform. They do not always behave as hard, brittle materials; given heat, confining pressure and sufficient time, rocks can bend, and they can flow. Depending on physical conditions and time, they exhibit different combinations of viscous and elastic behavior. The terms *viscous* and *elastic* have distinct meanings. When either a viscous substance or an elastic substance is subjected to a deforming force, it changes shape; however, when the force is removed the elastic one reverts to its original shape, but the viscous one does not. The same rock may deform in a brittle, elastic, viscous, or visco-elastic (delayed-recovery) manner depending upon the size of the differential stress, the confining pressure, the presence and nature of pore fluids, the temperature, and the strain rate—that is the rate at which the rock changes shape.

The words *stress* and *strain* are often incorrectly substituted for one another in ordinary speech. They have related but very different meanings, stress being force per unit area; and strain being deformation resulting from applied force. We may describe stress and strain in terms of triaxial ellipsoids—three-dimensional objects that look like flattened footballs—whose axes correspond to the directions and relative magnitudes of the greatest, smallest, and intermediate stresses or strains. The differential stress or strain is the difference between the greatest and the smallest stress or strain.

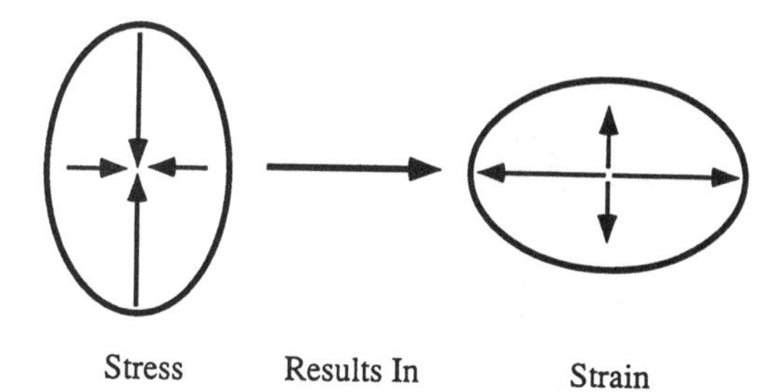

When rocks behave as viscous materials, they may create folds. Folds are wonderful! To see a great bed of massive sediment, which you know once lay flat on the seafloor, folded back upon itself across the face of a mountain, is to apprehend the titanic forces of *orogenesis,* or mountain building. The first scientist to notice such folds and write about them was a pioneer alpinist, Horace-Bénédict De Saussure (1740–1799). Born and educated in Geneva, De Saussure was a brilliant student, a professor of mathematics at twenty, and professor of natural philosophy at twenty-two. Above all, he loved the mountains, and by his eighteenth birthday he had climbed all of them near Zurich. As an adult he went farther afield and higher, and he went always as a *geologist*—a word he coined along with *geology*—systematically studying the rocks, structures, and landforms of his alpine playground. He published a detailed account of his studies, *Voyages dans les Alpes,* in four volumes between 1779 and 1796. During those same years he hiked across the Alps at least 14 times and made 16 other trips from the foothills to the central axis and back.

As a scientist, De Saussure's major ideas dealt with geomorphology and glaciers. It was he who first suggested that glaciers flow downhill in response to their weight and carry with them debris from higher elevations. Yet he failed to extend this insight to the erratic boulders of granite and schist on the southeastern flanks of the Jura Mountains, nearly fifty miles from present glaciers. Had he done so, he might have discovered the Ice Age. Local people, mountaineers, and others familiar with the terrain, recognized the evidence for elongated prehistoric valley glaciers long before this idea was accepted by the scientific community. That realization did not come to geologists for more than fifty years and then only slowly and with passionate debate. De Saussure's explanation for those boulders in that place was typical of his time: catastrophic flood waters associated with the biblical deluge.

De Saussure saw and sketched many of the great alpine folds, and he realized, towards the end of his career, that the folding was most likely due to some horizontal force, whose source he never identified. His contributions to geology were mostly descriptive, not theoretical; yet by providing an inspiring and enthusiastic description of the Alps, including fauna, flora, and native culture as well as geology, De Saussure began the long tradition of alpine geology that has attracted hundreds of young scientists to the high peaks.

There are several varieties of folds; the simplest are called *anticlines* and *synclines*. If you see them in a cross-section, such as a roadcut, they look like upright (synclines) or inverted (anticlines) *U*s.

When you look at folded rocks from an airplane, if there is topographic relief, you see that the more resistant layers form ridges, and the less resistant form valleys. Because many folds are tilted along their axes, the ridges eventually double back, creating a pattern that resembles side-by-side canoes. If you have difficulty visualizing this map pattern, you can make a model with an old magazine. Open the magazine completely at the middle (where the staples come through) and pretend that it is a pile of sedimentary rocks. Now close it back up so that it is a folded pile of sediments, and the spine is the axis of the fold, a syncline or an anticline as you choose. Hold the spine of the magazine along the edge of your kitchen table, which will represent the Earth's horizontal surface. Tilt the spine (the fold axis) and cut (erode) the magazine parallel to the table top. Voilà!

Eroded folds are very common in mountain ranges such as the Appalachians, where the combined effects of folding and differential weathering have produced belts of parallel ridges and valleys that stretch from the Hudson River valley to Birmingham, Alabama. Other spectacular examples are to be found in the Alpine Ranges that span Eurasia from the Pyrenees to Sumatra, the North American Cordillera, that extends from Alaska to Mexico, and numerous other mountain ranges around the world. Look for them, especially

Anticlines and synclines of the Jura Mountains.[64]

when you are traveling, by land or air, across the linear trend of a mountain system.

When rocks experience brittle failure, the resulting structures are *joints* or *faults*. Joints are fractures without slippage (or displacement) along the plane of the fracture. Faults are fractures along which displacement has occurred. Faults are classified according to the sense of the displacement. Compare normal and reverse faults as seen in a vertical section, such as a road cut:

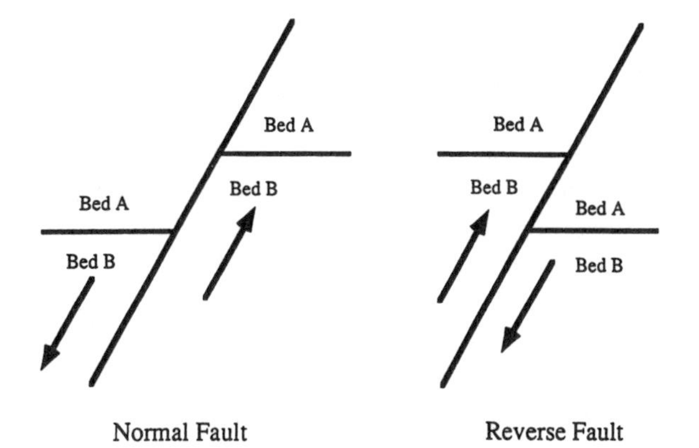

In the normal fault, the maximum stress is vertical and the minimum stress is horizontal. Notice that the displacement causes sideways extension, which means that normal faults are likely to occur where the crust is spreading sideways. In the reverse fault, the maximum stress is horizontal and the minimum stress is vertical. The displacement causes sideways contraction, which means that reverse faults are likely to occur where the crust is being squeezed sideways.

In normal and reverse faults the principal displacements are up or down. There are also faults along which the principal displacements are horizontal. These are called *transcurrent* or *strike-slip* faults, and for them the maximum and minimum stresses are in the horizontal plane. Two kinds of motion are possible, as shown in this map view. Imagine that you are standing astride the fault when an earthquake occurs. If the righthand side moves towards you, the fault is called *right lateral.* If the lefthand side moves towards you, the fault is called *left lateral.*

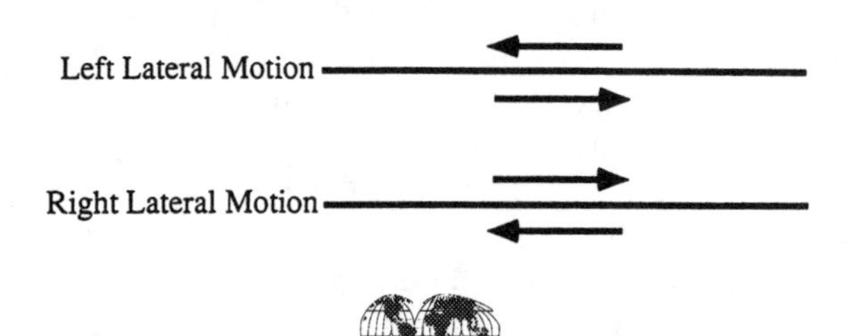

During the mid-1960s, when the *Theory of Plate Tectonics* was being developed, a professor of geophysics at the University of Toronto named J. Tuzo Wilson (1908–1993) had an extraordinary insight: A transcurrent fault connecting two offset sections of mid-ocean ridge would have a sense of motion just the opposite of that occurring along an ordinary transcurrent fault because of the creation of new crust at the mid-ocean ridge. Wilson named these, and all other faults that connect segments of accreting or consuming plate margins (spreading centers or subduction zones), *transform* faults. He also recognized that, although there might be a scarp (cliff) or some other surficial indication of the fault line beyond the spreading centers, the active part of the fault is restricted to the part between the spreading centers.

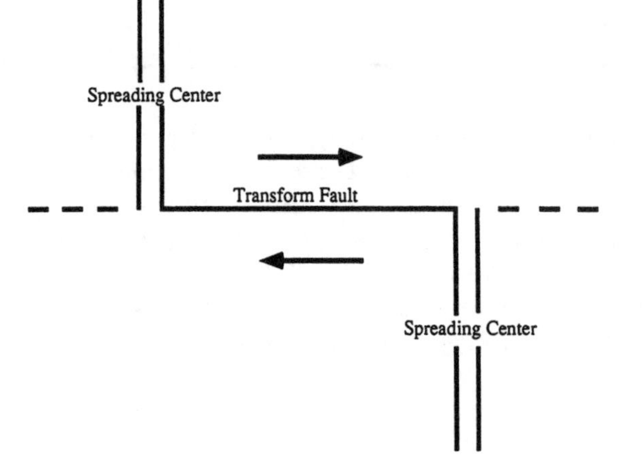

At this point I am fairly certain that you have not understood my description of transform faults. That is because they are almost impossible to understand by simple reading or listening. I remember the evening, probably in 1966, when I heard Tuzo Wilson describe transform faults to a large audience at Rutgers University. He had brought a stack of heavy colored paper and an enormous pair of shears onto the stage, and as he talked he cut the paper and attempted to demonstrate the curious, anti-intuitive motion of faults that connect spreading centers. His attempt failed, at least for me. The papers were flying about on the stage in comic confusion, but I really didn't understand what he was telling us. Later I demonstrated the phenomenon to myself, with paper and shears, and then I understood it. You will understand, too, if you take paper and shears and do the following:

1. Draw straight lines A-B, B-C, and C-D. Write the letters on the paper and also draw the short line that crosses C-B at a right angle.

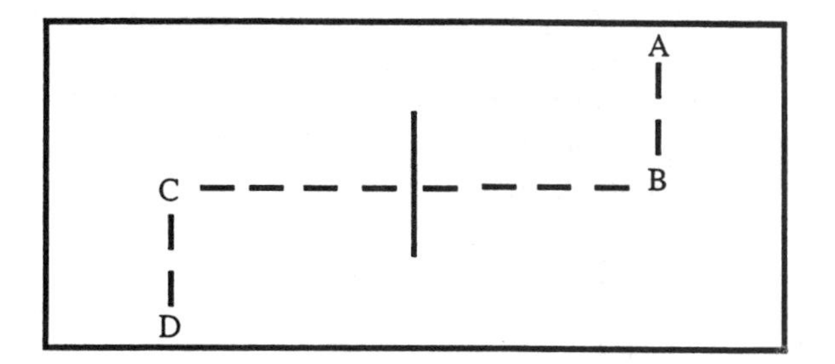

2. Cut the paper in two precisely along the lines A-B, B-C, and C-D.
3. Pretend that A-B and C-D are offset sections of a mid-ocean ridge and that C-B is a transcurrent fault connecting them. Fit the two halves of the paper together on the table and slowly pull them apart so that A-B and C-D (but not C-B) open up, like a spreading mid-ocean ridge. What is the sense of motion on the fault C-B? That's what Tuzo Wilson discovered.

Within three years of Wilson's suggestion, in one of the most spectacular confirmations of the Theory of Plate Tectonics, three geophysicists at Columbia University—Brian Isacks, Lynn Sykes, and Jack Oliver—found exactly the kinds of earthquake activity Wilson had predicted for ridge-to-ridge transform faults in the North Atlantic.

Most transform faults are out of sight, beneath the oceans, but a few can be seen on the continents. The great example in the United States is the San Andreas fault of California, which separates the North American and Pacific plates. The San Andreas is a right lateral fault that moves, on average, 4 to 6 centimeters each year. Because their movements cause powerful earthquakes in a densely populated region, the San Andreas and associated faults are subjects of continual study and concern.

Another class of great faults in the continental crust are regional thrust belts. These are belts of parallel, low-angle, reverse faults that form in response to the horizontal, compressive stresses associated with converging lithospheric plates. The overthrust sheets are beds of sedimentary rocks, which may be hundreds of miles in extent yet only thousands of feet thick. They slide along layers of weak materials, such as shale, without deforming the underlying rocks. Occasionally a fault will turn abruptly upwards, cut across stronger beds, and then continue its nearly horizontal movement along another weak layer, higher in the section.

Regional thrust belts may reduce the original horizontal extent of the displaced sedimentary rocks by as much as 50 percent. How such thin sheets can be transported so far, at such low angles, has been a mystery from the time of their discovery, but two factors, gravity and fluid pressure, may provide the answer. In 1959 two American geologists, M. King Hubbert and William Rubey, calculated the forces required to move thrust sheets of the shape and size known to exist along the Rockies and along the Appalachians. They compared those forces with the known strength of rocks and concluded that the thrust movements were physically impossible unless the effective weight of the thrust sheets could be significantly reduced. They then suggested that this effective weight would be significantly reduced if the rocks of the thrust sheet contained high-pressure fluid. The greater the fluid pressure, the smaller the force required to move the sheet. With sufficiently high fluid pressure, the forces required to move the sheet would be within the strength limits of the rocks. The role of gravity is to distribute the forces throughout the moving sheet rather than to apply a push at the back or a pull from the front. This may be an important factor in some thrust events but not in all, given that some thrust sheets have moved uphill!

Although gravity does not explain every structural detail of Earth's crust, it is the dominant force, and all terrestrial phenomena must deal with it. Whatever the actual mechanisms, forces due to gravitational attraction drive the tectonic plates and cause most of the strain observed in rocks.

Isaac Newton presented his *Law of Universal Gravitation* in 1686. In modern language the law states: "Every particle of matter in the universe attracts every other particle with a force which is directly proportional to the product of the masses of the particles and inversely proportional to the square of the distance between them."[65]

$$F = G \times [\text{mass}_1] \times [\text{mass}_2] \div [\text{distance}]^2$$

In the equation stating Newton's Law of Universal Gravitation, the proportionality constant, G, known as "big G," is the universal gravity constant. The force of gravity is thought to be one of the basic forces of the universe. Like the velocity of light, G is believed to be a fundamental constant, although there has been recurrent speculation that it may vary over time.

Newton's Law of Universal Gravitation is an example of an inverse square law. Another example is furnished by a light bulb. The intensity of light coming from a light bulb decreases as the square of the distance from the bulb increases. Why? The intensity of light is distributed evenly over the surface of a sphere surrounding the source; moving out from the source, the light is diluted as it is spread over the sphere's increasing surface, which grows as the square of the radius (Area = $4\pi r^2$). Inverse square relationships obtain whenever a phenomenon has a point source and extends out equally in all directions.

The weight of an object on Earth, such as you, is the force of gravitational attraction between that object and Earth:

$$\text{Your weight} = G \times [\text{your mass}] \times [\text{earth's mass}] \div [\text{earth's radius}]^2$$

Your weight also equals your mass times the acceleration due to gravity on Earth, which is called "little g" (to distinguish it from "big G").

$$\text{Your weight} = g \times [\text{your mass}]$$

So,

$$g \times [\text{your mass}] = G \times [\text{your mass}] \times [\text{Earth's mass}] \div [\text{Earth's radius}]^2$$

$$\text{Therefore, } g = G \times [\text{Earth's mass}] \div [\text{Earth's radius}]^2$$

Inasmuch as big G and Earth's mass are constant, variations in little g are due to changes in the distance to the center of the Earth, which varies with latitude and elevation. This is a major effect, but we must also consider that Earth's density is not distributed uniformly. The force of gravity will be greater if the rocks buried beneath our feet are more dense than normal. The converse is also true, of course, so if you want to weigh less without dieting, go to the equator and stand on the highest mountain made of low-density rocks such as shale!

We can measure g in the field. By applying corrections for elevation and latitude we can make educated guesses about the densities of the rocks beneath our instrument. Sometimes guesses like this can locate bodies of dense ore minerals, or they may suggest possible geometries for deep structures.

Among his many elucidations, Isaac Newton showed that a rotating planet should be an oblate spheroid rather than a sphere, flattened slightly at the poles and bulging a bit at the equator. To test this idea the French Académie des Sciences sent surveying parties to Ecuador and to Lapland, in 1735 and 1736, to measure the distances between polar and equatorial lines of latitudes. Newton's calculations were confirmed—the polar diameter of the Earth is 12,714 kilometers, and the equatorial diameter is 12,756 kilometers.

The leader of the French expedition to Ecuador was Pierre Bouguer. In 1738, while in the vicinity of a 20,000 foot volcano called Chimborazo, Bouguer observed that the mountain was attracting his plumb bob sideways.* He also noted that the amount of deflection was less than predicted by Newton's Law. The wintry conditions the expedition experienced in the high Andes spoiled any chance Bouguer and his colleagues had of making precise measurements of these effects, but their observations inspired Nevil Maskelyne, Astronomer Royal of England, to convene The Committee of Attraction in 1772, under the auspices of the Royal Society of London. Among its members was Henry Cavendish, one of the most brilliant and wide-ranging scientists of the eighteenth century.

The Committee sent Maskelyne to Scotland in July 1774 to measure the deflection of a plumb bob by a 3,500 foot mountain in Perthshire, named Schiehallion. Maskelyne measured the deflection, and thereby weighed the Earth. His technique was to use a "pan balance," such as Iustitia (or Lady Justice) carries, with Earth on one pan and the mountain on the other. The indicator for the balance was the plumb bob.

At two locations, on opposite sides of the mountain, Maskelyne established the angular deflection of the plumb bob away from vertical by observing stars. He then estimated the mass of the mountain

*A plumb bob is the metal weight on the end of a plumb line, which is the traditional tool for determining verticality.

from its surveyed volume and composition. With these data, Maskelyne could calculate the mass of the other attracting body, the whole Earth.

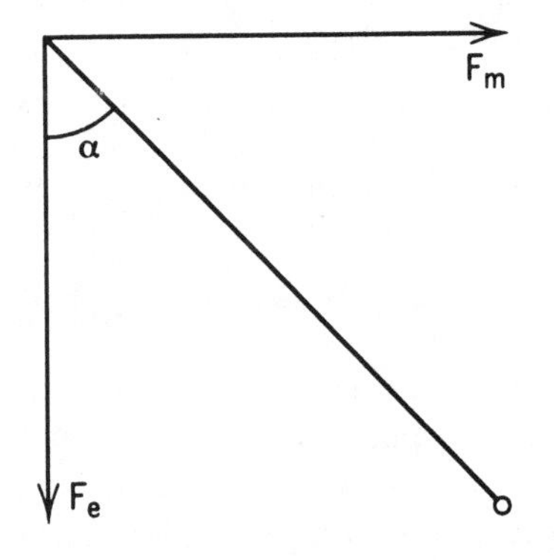

The method works like this. The forces on the plumb bob can be represented as vectors:

F_e is the weight of the plumb bob.

F_m is the sideways force between the mass of the mountain and the plumb bob.

α is the angle of deflection of the plumb line from vertical in degrees.

M_b is the mass of the plumb bob.

M_e is the mass of the Earth.

M_m is the mass of the mountain.

D is the distance between the plumb bob and the mountain.

R is the radius of the Earth.

According to Newton's Universal Law:

$$F_e = G \times M_b \times M_e \div R^2$$

The geometry of the vector triangle shows that

$$F_m = F_e \times \text{Tan}\alpha = G \times M_b \times M_e \times \text{Tan}\alpha \div R^2$$

It is also true that

$$F_m = G \times M_b \times M_m \div D^2$$

Equating these two expressions for F_m gives

$$G \times M_b \times M_e \times \text{Tan}\alpha \div R_2 = G \times M_b \times M_m \div D^2$$

After eliminating G and M_b from both sides of the equation, we solve for M_e, the mass of the Earth:

$$M_e = M_m \times R^2 \div D^2 \times \text{Tan}\alpha$$

Everything on the righthand side of the equation can be measured directly. Maskelyne observed a mean deflection of his plumb bob of about 11.5 minutes of arc.* He calculated that the average density of the Earth was 4.5 times the density of water and almost twice the density of normal, crustal rocks.

Maskelyne's estimate was 20 percent too low and also lower than expected. Newton himself had stated that the average density of Earth would be five to six times that of water.[66] Nevertheless, Maskelyne's attempt was the first experimental demonstration that the deep Earth contains much denser materials than the crust.

After spending July through October in a small tent on the mountain, Maskelyne and his assistants were sufficiently pleased with their efforts to host a celebration. A keg of whisky was brought up from the nearby village of Kinloch Rannoch, and many of the local folk invited. A fine party it was, to be sure, for they burned down the tent, including the cook's fiddle. Maskelyne replaced the fiddle and was remembered by the grateful cook, red-haired Duncan, in a Gaelic song. Derek Howse, Maskelyne's biographer, has provided a translation of the pertinent stanzas and refrain:

The Yellow London Lady

On the trip I took to Schiehallion
I lost my wealth and my darling;
There was not her like in Rannoch
When she stirred herself to song

*Eleven and a half minutes are between one-fifth and one-sixth of a degree.

Though before I was mournful
This year I am joyful
I have received the yellow London Lady
and she is the instrument for music.

It was in Italy that she was reared;
And she came across the sea to London,
To give music to king and queen
And to the high gentry of Europe.

She came of a precious family
From the company of genteel royalty,
From the high astronomer of the kingdom
Who paid for her with his gold.

It is Mr. Maskelyne, the hero,
Who did not leave me long a widower,
He sent to me my choice treasure
That will leave me thankful while I live.[67]

The principal difficulty with Maskelyne's experiment was uncertainty about the mass of the mountain and of the material beneath the mountain. This uncertainty was appreciated by Henry Cavendish and the Committee of Attraction. In 1798 Cavendish published the results of a laboratory experiment that eliminated the geological uncertainties by using lead spheres of precisely known mass. The *Cavendish Experiment* determined that the Earth's average density is about 5.5 times the density of water. The currently accepted value is 5.517 grams per cubic centimeter.

Bouguer and Maskelyne demonstrated, as Newton had earlier predicted, that surveyors might have a problem making accurate surveys in regions where large mountains were present. This was a particularly timely concern because over the next seventy-five years the British were to make a precise survey of the Indian subcontinent, virtually in view of the vast mass of the Himalayas. Accordingly, the British surveyors prepared to correct for the gravitational attraction of the Himalayas on their plumb bobs; however, they were surprised to find that their surveys were constantly in error. Apparently the plumb lines were not deflected as much as the surveyors had expected. On the Ganges plain, 56 miles from the feet of the Himalayas at a place called Kaliana, the plumb-line deflection was only 1 second of arc or 1/60 of a degree; they had calculated that it should have been nearly 1 degree! After checking their estimates for the visible mass of the Himalayas, the surveyors realized that a mass deficiency must lie beneath the mountains. The mountains must have low-density roots.

In order to understand this you need to understand the principle of Archimedes (287–212 B.C.), which, according to legend, the Greek savant discovered while sitting in the public baths of Syracuse pondering a problem posed by King Hieron. The king wanted to know how much silver had been secretly substituted for gold in a wreath he bought. As everyone knows, Archimedes raced home from the baths, naked as a flea, crying "Eureka!" (I've found it!) to his astonished neighbors. He had found a way to compare the densities of gold, silver, and the suspect wreath; more important, he had discovered that a fluid exerts an upward force on an immersed body that is equal to the weight of the fluid displaced.* In the Earth the mantle behaves like a fluid and, like a boat in water, the rocks of the mountain, which are less dense than the mantle, displace enough of the mantle to support the overlying mountain. What this means is that the Himalayas and other mountain ranges are floating. Like icebergs in the sea, mountains on Earth sink into the mantle until they reach buoyant equilibrium. The higher the mountain range, the deeper its roots. As mountain tops erode, the root zones float up towards the Earth's surface. Old, deeply eroded mountain ranges

---

*Consider the forces on a spherical surface within a glass of water. At each point the force on the surface depends only upon the depth of the water at that point. Because the water is at rest, the sum of the forces on the spherical surface must equal the weight of the water inside the sphere. If we replace the water inside the sphere with an identical volume of some other material, what will happen? The forces exerted on the sphere by the surrounding water will not change. If the new sphere weighs less than the sphere of water, it will float. If the new sphere weighs more, it will sink. If the new sphere weighs the same as the sphere of water, nothing will happen.

have shallow roots; young, high mountains have deep roots. This condition, which is called *isostatic equilibrium*, distinguishes Earth from other rocky bodies in the solar system.

The rising of low-density roots, or *isostatic rebound*, is too slow to notice when caused by erosion. Noticeable isostatic rebounds occur when glaciers melt, suddenly removing the load of several miles of ice from the crust. In North America after the latest glaciation, for example, the crust rebounded hundreds of meters, rising at times more than a centimeter per year. The subsurface adjustments associated with isostatic rebound are typically accomplished by rocks that are so deep and hot that they flow quietly, without accumulating the elastic strain energy that colder, stronger rocks accumulate and release in earthquakes.

Seven

On Trembling Ground in an Ambivalent Field

In August of 1875 one of the great Victorian adventurers, John Milne (1859–1913), departed England for Japan to begin his appointment as Professor of Geology and Mining at Tokyo's Imperial College of Engineering. Upon his arrival seven months later, after crossing Europe, Siberia, Mongolia, China, and the China Sea, he was rolled out of bed by an earthquake, whereupon earthquakes became his passion. Milne documented the locations and effects of earthquakes; he designed and built new instruments to detect and measure earthquakes; he founded the Seismological Society of Japan; and in 1883, at a Society luncheon, he made a remarkable suggestion: "It is not unlikely," said Milne, "that every large earthquake might, with proper instrumental appliances, be recorded at any point on the surface of the globe."[68] That earthquake waves might shake the other side of the planet may have seemed farfetched, but Milne was right. On March 22, 1894, pendulums in Germany recorded earthquake waves that had, in fact, passed through the Earth from Japan.

Most earthquakes are caused by movements on faults within the lithosphere, the outer 100 kilometers of the Earth where the rocks are cool enough and therefore strong enough to accumulate elastic strain energy until they break. As rocks are deformed, elastic strain energy is stored in them, just as it is stored in a stretched rubber band, a drawn bow, or an inflated balloon. At some point the strain energy is sufficient to overcome the frictional forces holding the rock together, and the rock breaks. The elastic strain energy is suddenly transformed into elastic waves that move rapidly out from the point of rupture or, as it is called, the earthquake focus. Sooner or later these waves are themselves transformed into heat, sound, and mechanical work.

Deep earthquakes, which range from 100 to 680 kilometers down, are associated exclusively with subduction zones. This is because rocks are such poor conductors of heat that as the lithosphere descends into the planet it remains colder and stronger than the surrounding mantle, which is too hot and weak to accumulate elastic strain energy and to rupture. In subduction zones earthquakes due to rupture extend to depths of about 300 kilometers. Below this depth they are less frequent and more mysterious. Below 400 kilometers they may be due to a crystal structure change that transforms ordinary olivine into a more dense (spinel-like) form.* This olivine-to-spinel transformation, however, does not appear to explain the very rare, very deep, very powerful earthquakes. The biggest deep quake ever recorded released more energy than 240 million tons of exploding TNT 670 kilometers below Bolivia on June 9, 1995. Such quakes may be due to another crystal structure change, or they may be due to a still unknown mechanism.

The first known seismograph was designed around A.D. 132 by Chang Heng, Astronomer Royal in the Later Han Dynasty. His instrument, which was a work of art as well as advanced technology, was a large bronze jar, six feet in diameter and perhaps eight feet high, with eight, regularly spaced dragon's heads poking out around the top. Below each dragon, on the floor around the jar,

*The spinel structure, named for the mineral spinel ($MgAl_2O_4$), has oxygen atoms packed as closely as possible and metal atoms stuffed into interstices between either 4 or 6 oxygen atoms. Spinel itself has magnesium atoms surrounded by 4 oxygens and aluminum atoms surrounded by 6. When olivine (Mg_2SiO_4) is subjected to sufficient pressure its more open, low-pressure structure collapses into a spinal-type structure and magnesium atoms surrounded by 6 close-packed oxygens and silicon atoms surrounded by 4.

were eight bronze toads, looking upwards expectantly with their mouths open. Each dragon carried a ball in its mouth. The first earthquake wave caused a mechanism within the jar to nudge one dragon, which then dropped its ball into the gaping mouth of the toad below. The loud clang alerted the Emperor and his government that an earthquake had occurred, and by recovering the ball they learned in which direction the affected region lay.

> On one occasion [according to the official historian] one of the dragons let fall a ball from its mouth though no perceptible shock could be felt. All the scholars at the capital were astonished at this strange effect occurring without any evidence of an earthquake to cause it. But several days later a messenger arrived bringing news of an earthquake in Lung-Hsai [400 miles away]. Upon this everyone admitted the mysterious power of the instrument.[69]

Seventeen centuries later the Emperor of Japan and his government encouraged John Milne and his colleagues to provide similar information. They and other nineteenth-century students of earthquakes developed seismometers and seismographs with which to measure and record earthquakes with increasing precision and sensitivity. These instruments are of two kinds: inertial devices and strain gauges.

Chang Heng's bronze jar was an inertial device. An inertial seismometer contains a massive object that is suspended or supported so that it will not be affected when a wave passes through the Earth. The suspended mass, which may be still (a weight on a spring) or moving (a pendulum), provides a stable reference with which to compare the motion of the planet. You can build a crude inertial seismograph by suspending a massive object, such as a stack of barbell weights, above the ground with a garage door spring. Ground vibrations can be observed (in a dark room) by attaching a penlight to the weights and shining it through a pin hole onto the wall. Unless you live in a very remote location, you will find that your everyday world quivers with vibrations from many sources.

The strain gauge measures distortion of the Earth and is extremely sensitive. To understand how it works, imagine a guitar string stretched between two posts in the ground; a change in the pitch of the plucked string would indicate that the ground had moved. Such devices are used to monitor strains in the Earth—for example, along the San Andreas fault—or in manmade structures such as bridges or skyscrapers.

By the time seismic waves were being detected through the Earth, a substantial body of theory was available for their interpretation. As we discussed earlier, when contrasting the particulate theory of Newton and the undulatory theory of Huygens, physicists and mathematicians of the early nineteenth century studied the transmission of light through crystals. In 1821 a young French engineer, Augustin Fresnel (1788–1827), observed that certain polarized-light beams passing through a crystal did not interfere with one another, leading him to suggest that light travels through crystals as transverse (sideways) rather than longitudinal (forward and backward) vibrations. Fresnel's hypothesis attracted the attention of Siméon Denis Poisson (1781–1840), a French mathematician, who determined that an elastic solid could simultaneously transmit both longitudinal and transverse waves and that the longitudinal waves would travel almost twice as fast as the transverse waves.

Fifty years later, when German pendulums detected the Japanese earthquake of March 22, 1894, their pens marked the arrivals of both the longitudinal and transverse vibrations. Although this was not immediately recognized, it wasn't long before R. D. Oldham (1858–1936) of the Geological Survey of India recognized the different wave signals in seismograph records.

Longitudinal body waves are called *compressional* or *P* waves (for primary, because they arrive first). They are the fastest seismic waves, with velocities ranging from more than 3,000 miles per hour in water to more than 29,000 miles per hour at the base of the lower mantle.* P waves can pass straight through the Earth in about 20 minutes. While one passes through a substance (liquid or solid), the particles move back and forth, parallel to the direction of propagation, and the substance is momentarily compressed. In this way, P waves are like sound waves.

---

*Velocities are normally stated in kilometers per second: 1 km/sec = 2,237 mph. Thus P wave velocities range from 1.3 km/sec in water to 13 km/sec in the lower mantle.

Transverse body waves are called *shear* or *S* waves (for secondary, because they arrive after the P waves). Their velocities vary from almost 7,000 miles per hour in granite to almost 16,000 miles per hour at the base of the lower mantle. S waves will not pass through a liquid. When one passes through a rock, particles in the rock move at right angles to the direction of propagation. In their transverse motion, S waves resemble light waves.

During a thunderstorm the lightning flash travels much faster than the thunder sound and arrives first at any point beyond the event. The farther they travel, the greater the delay between the light and sound waves. The same is true of P waves and S waves in the Earth. Just as we use the time delay between light and sound to estimate the distance to a lightning bolt (approximately two-tenths of a mile per second), Oldham and his colleagues used the delays between the P and S wave arrivals to determine the distances from their instruments to the earthquake locations. It was then possible to map the distribution of tremors relative to their point of origin, and this led Oldham in 1906 to a surprising observation: Records from seismographs up to 120° (1,452,000 kilometers) away from earthquake locations showed first wave arrivals at times that increased regularly with distance from the quake; beyond 120°, however, P waves were abnormally delayed and S waves were either absent altogether or (he supposed) severely delayed. Oldham attributed this abrupt change in wave transmission to a spherical region of reduced seismic velocities in the center of the planet; he had discovered the Earth's core.

Oldham's discovery was based upon a law of optics attributed to Willebrod Snell (1591–1626) in 1621. According to Snell's Law, when a wave passes from a fast medium into a slower medium, the direction of the wave is refracted towards a line drawn perpendicular to the interface, and when a wave passes from a slow medium into a faster medium the direction of the wave is refracted away from the perpendicular. This phenomenon of wave refraction is well known to kids at summer camp. Spearing fish from a boat requires an adjustment for the refraction of light waves. The light that carries the image of the fish to the hunter travels from a slower medium, the water, into a faster medium, the air. Unless the fish is directly below, the light is bent away (down) from the vertical, so the fish appears above it's actual position. If you want fish for supper, get directly above your target or aim low.

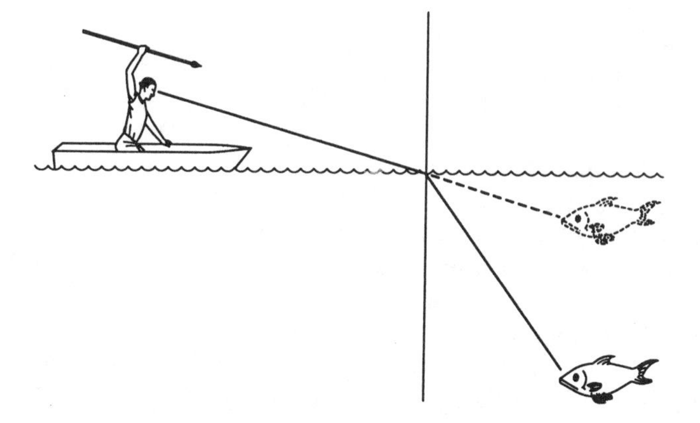

Seismic wave velocities tend to increase with pressure (depth) unless the nature of the material changes. Therefore, as seismic waves descend into the Earth they tend to be refracted away from the vertical. Superimposed upon this general trend are dramatic velocity changes caused by changes of the earth materials. The presence of a little bit of molten rock in the asthenosphere may cause the observed slowing of velocities there. Moreover, abrupt crystal structure changes may cause sudden velocity changes within the deep mantle. The greatest change of all is at the mantle/core boundary, where P wave velocities decrease sharply and S waves stop altogether. Using Snell's Law, you can see, as Oldham indicated, how a sphere of slower material within the Earth would create the uneven distribution of wave signals he observed.[70]

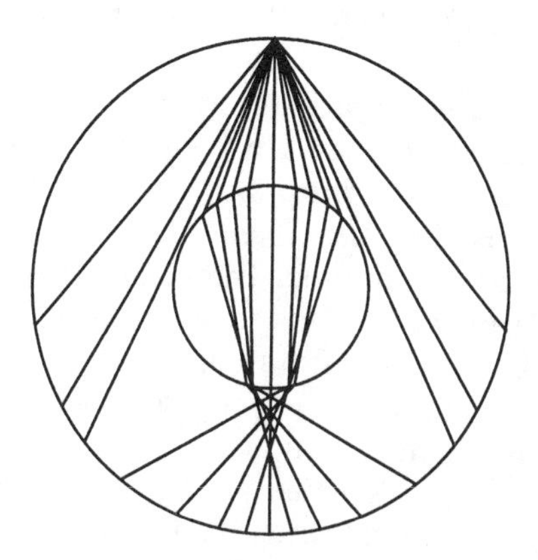

The discovery of a second great discontinuity within the Earth also came early in the twentieth century. In 1909, Yugoslavian seismologist Andrija Mohorovicic (1857–1936) discovered that seismic velocities increase abruptly 54 kilometers below the surface of southeastern Europe, which he attributed to a fundamental change in the rocks at that depth. Mohorovicic's discontinuity was subsequently identified all around the planet. The Moho, as it is now called, separates the Earth's crust from the underlying, faster (and more dense) zone called the mantle. Its depth varies from place to place, but average values are 40 kilometers beneath continental regions such as the Canadian Shield and 15 kilometers beneath the ocean basins.

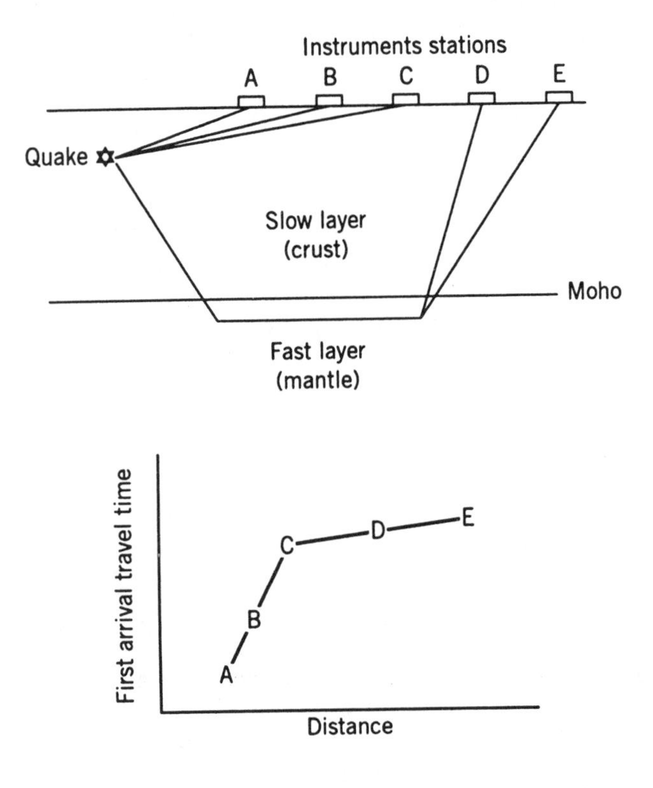

Mohorovicic's discovery was based upon observations of the same earthquake from several seismographs located at different distances from the quake site. When he plotted the travel times for the first P wave arrivals against the distance, he found that different relationships obtained for nearer and farther stations. The explanation was that it was faster for a P wave to reach a distant station by dropping down into the higher-velocity mantle than by going the shorter route through the slower crust. An analogy can be made with automobile travel across a city: For short trips the direct route is probably fastest, but for long trips it may be faster to go out to a high-speed beltway and drive around the city.

Through the decades since Mohorovicic and Oldham discovered the crust/mantle and mantle/core discontinuities, seismology has made great progress. Three technological advances have been crucial. First, instrument design was improved, the number of instruments was increased, and techniques were developed to use arrays of instruments. Major motivating factors for this advance were our interest in monitoring nuclear weapons testing in the Soviet Union, our commercial interest in deeper oil and gas exploration, and our scientific interest in seismic activity on the Moon and Mars. Second, the creation of atomic clocks and satellite communications systems improved our ability to determine arrival times precisely and consistently; the satellites offer the additional advantage of creating global instrument networks that can be in constant communication with each other. Third, digital technology and computers have revolutionized our ability to collect, organize, and interpret vast bodies of data.

Revolution is an overworked word, especially in twentieth century science, but no other word will do. Ours *is* a new world. In seismology, more than any other subdiscipline of earth and planetary science, we are at last able to execute experiments and observe Nature on scales appropriate to the subject. For me, this realization inspires a Christmas Eve-like anticipation of the great discoveries that lie ahead, and a profound respect for Milne, Oldham, Mohorovicic, and other pioneers who discovered so much with so little.

The patterns of seismic waves passing through the Earth and recorded by many seismographs can be generally accounted for by a model of concentric spherical shells, beginning with the crust and ending with a small, solid sphere at the center of the planet. The depths of the main shells are described as follows in a scheme proposed by Keith E. Bullen (1906–1976) in the 1930s and brought up to date by Bruce A. Bolt in 1982:[71]

A	Crust	0–40 km
B	Noncrustal lithosphere (upper mantle)	Moho to 100 km
C	Asthenosphere	100–640 km
D′	Lower mantle	640–2,780 km
D″	Transition shell	2,780–2,885 km
E	Outer core	2,885–4,590 km
F	Transition shell	4,590–5,155 km
G	Inner core	5,155–6,371 km

Within or between these shells major changes in seismic velocities occur in several relatively thin zones. First is the *Mohorovicic Discontinuity,* which marks an abrupt increase in seismic velocities within the lithosphere and is generally attributed to a change in composition from the less dense, slower rocks of the crust to more dense, faster rocks of the upper mantle. At the base of the lithosphere, there is a discontinuous region of slower transmission, between 100 and 250 kilometers down, which may be a zone of partial melting. Next is a zone of rapidly increasing velocities, between 300 and 400 kilometers down, which may correspond to structural phase changes among the silicate minerals, such as the olivine-to-spinel change mentioned earlier. Next, around 700 kilometers down, is another zone of rapidly increasing velocities, where more crystal structure collapses occur. This is the top of the lower mantle, which is the largest of Earth's major shells. Experimental studies indicate that the dominant mineral in the lower mantle is an iron-magnesium silicate with the elemental proportions of a pyroxene (i.e., $MgSiO_3$), but one with a much more dense crystal structure involving close packing of oxygen, iron, and magnesium atoms with silicon atoms surrounded by six—instead of four—nearest neighbors. Considering the vast size of the lower mantle, this exotic mineral, called magnesium silicate perovskite, is probably Earth's most abundant mineral. At the base of the lower mantle, 2,900 kilometers below the surface, the Earth changes from dense rock to molten iron. P-wave velocities decrease suddenly, and S waves disappear. The final, deepest sudden change is at the outer/inner core boundary where P wave velocities increase upon entering a solid region.

The model just described is based upon a simple series of nested, spherical shells. It accounts for gross characteristics of the seismic data, but other data indicate that the Earth's interior is more complex. The most exciting indications come from a technique called *Seismic Tomography.*

Tomography is a technique developed by medical radiologists in the 1930s in which an X ray tube and an X ray detector are rotated around a patient in order to obtain an image of a specific plane. The advent of computers brought the improvement called CAT scanning (computerized axial tomography), which is a mainstay of modern medical diagnosis.

Tomography can be practiced with other kinds of radiation. In the 1970s, as the quantity and quality of their data increased, geophysicists began using tomographic techniques to define lateral seismic velocity differences within the Earth. Seismic tomography is able to identify regions within the planet where seismic waves are relatively fast or slow. Fast regions are thought to be colder, slow regions hotter. Therefore, the patterns of lateral variation may show the distribution of hotter and colder materials within the planet. Such patterns would probably tell us the shape of convection cells and other thermal features such as foundered slabs of cold lithosphere, plumes of hot material, or down draughts of cold materials.

One inconvenience associated with seismic tomography is that, except for the extremely heterogeneous region immediately above the mantle/core boundary, thermal homogeneity increases with depth; that is, shallow regions of greater complexity obstruct our view of the simpler, deeper levels. In spite of this and other problems, however, great progress has been accomplished over the last decade. The areal resolution in 1992 was 5 by 5 degrees, which equals an area approximately the size of Oregon on a spherical surface at the top of the lower mantle.

At this resolution what do we see? First, no layer appears to be homogeneous or isothermal, but the patterns are not simple; they do not resemble the simple convection-cell patterns one sees in a cup of coffee or a geology textbook. Second, at all depths the hemisphere centered upon the Pacific Ocean contains more low-velocity (hot) material than anywhere else in the planet; this is where most of Earth's heat has been for at least 500 million years. Third, down to at least 350 kilometers the ancient continental interiors of Canada, Siberia, Australia, Guayana, and Africa are underlain by high velocity (cold) materials; the lithosphere is very thick beneath these ancient surfaces. Fourth, slow (hot) materials underlie midocean ridges down to 100 kilometers, but the correspondence begins to weaken at 200 kilometers and disappears by 400. Fifth, patterns from as deep as 400 kilometers include small round hot spots, but these do not line up with crustal hot spots and therefore do not now support the Mantle Plume Hypothesis. Sixth, in the lower mantle the most prominent feature is a ring of high-velocity (cold) material that encircles the Pacific. This donut-shaped region is thought to be an accumulation of foundered lithospheric materials from the circum-Pacific subduction zones.

Seismic tomography is still young and unrefined, but it is exciting because, for the first time, we can see what the Earth's interior really looks like. One of the pioneers of seismic tomography, Don L. Anderson of the Seismological Laboratory at the California Institute of Technology, predicts that "in about 10 years [2005] we should have three-dimensional maps of the structure of Earth's interior, from surface to center, with a resolution of a few hundred kilometers. . . . We will see convection currents and shadows of past continental positions."[72]

Seismic tomography and other techniques, such as detailed analysis of the seismic wave forms and ultra-high pressure and temperature experimental techniques, are used to explore and define the nature of the core/mantle boundary region. Seismologists can now detect inhomogeneities a few tens of kilometers across, and experimentalists can now study chemical reactions at any temperature and pressure condition in the planet. The principal instrument in these extraordinary experiments is the diamond cell, a conceptually simple device in which the substance to be studied is squeezed between two diamond crystals and simultaneously heated by a laser. Results thus far suggest that the core/mantle boundary region is a 100 to 400 kilometer-thick zone of vigorous chemical reactions between molten iron from the outer core and dense solid oxides from the lower mantle. According to Raymond Jeanloz and Thorne Lay, two active students of the subject, "lower mantle rock has been and still is slowly dissolving into the liquid metal of the outer core."[73]

The varying state of the core/mantle boundary region has significant consequences for several phenomena observed at the surface, including heat flow and variations of the Earth's magnetic field.

Magnetism has been observed for a long time. The magnetic properties of magnetite, or lodestone, were known to the ancient Greeks; in the first century B.C. Chinese soothsayers constructed a magnetic compass to foretell the future; and by the fourteenth century A.D. the Chinese were using floating magnetic needles to navigate.

Brought to western Europe by Marco Polo (1254–1324), the technology of compasses guided the great fifteenth- and sixteenth-century European voyages of discovery. In 1600 William Gilbert (1540–1603), physician to England's Queen Elizabeth, published *De Magnete,* the first major attempt to describe and explain terrestrial magnetism. According to Gilbert, who experimented with miniature magnetic models of the planet adorned with tiny compass needles, Earth was itself a huge magnet. We know now that this is not the case, but Gilbert's concept prevailed for centuries.

624–565 B.C.	Thales of Miletus observes magnetic repulsion and attraction.
A.D. 720	Buddhist monks in China observe magnetic declination; begin first secular variation studies.
1267	Petrus Peregrinus discovers lines of equal magnetic force and their convergence upon points, which he calls poles.
1576	Norman discovers magnetic inclination and suggests that a north-seeking needle points not at the heavens but at the Earth.
1600	Gilbert publishes *De Magnete.*
1838	Gauss determines that Earth's field is an axially centered dipole, originating within.

1849	Delesse observes rocks magnetized parallel to Earth's field.
1895	Pierre Curie discovers upper thermal limit to magnetism, known today as *Curie temperature.*
1899	Folgheraiter suggests that thermal remanent magnetism in fired brick should parallel Earth's field.
1906	Brunhes speculates that Earth's field may have reversed in the past.
1926	Mercanton suggests possibility of worldwide record of field reversal.
1929	Matayuma observes reversed polarity in Manchurian lavas.
1951	Hospers observes alternating polarities in Icelandic lava sequences.
1960	Cox, Doell, and Dalrymple document magnetic polarity time scale for past 5 million years.
1963	Vine and Matthews observe bilateral symmetry of seafloor magnetic anomalies.

The Earth's magnetic field is a dipole whose lines of force are disposed as though there was a giant bar magnet inside the planet. The actual cause of the field is thought to be motion of extremely hot, metallic liquid in the outer core. On the surface we measure the direction of the magnetic lines of force. This is actually a three-dimensional situation: The measurement includes a vertical component, usually ignored, that is called the *magnetic inclination,* and which varies with latitude. At the north magnetic pole the inclination equals 90 degrees and the north-seeking needle points straight down. The inclination equals 90 degrees at the south magnetic pole also, and the north-seeking needle points straight up. Along the magnetic equator the inclination equals zero, so the needle is horizontal.

The axis of the magnetic dipole and the Earth's axis of rotation seldom correspond exactly. The angular difference between them is called the *magnetic declination,* which changes according to the longitude. To visualize this imagine walking around the base of a hill that has two poles stuck into its summit, one vertical and the other tilted. As you circle the hill the angular relationship of the two poles will change from perfect alignment at two locations to maximum disalignment at two locations. It also changes with time. At San Antonio in 1992 the declination was 9.5 degrees east of true north.

Why was William Gilbert wrong? How do we know there isn't a giant bar magnet buried in the planet? Because any magnetic solid loses its magnetism when heated above a certain temperature, known as the *Curie temperature.* Curie temperatures for common minerals are less than 700°C, so rocks deeper than about 50 kilometers cannot be magnetic.

Many minerals, especially those that contain large concentrations of iron, are ferromagnetic—that is, they behave as magnets. In rocks, these tiny crystal magnets may tend to point to where the north magnetic pole was when the rock formed. This phenomenon is known as *remanent magnetism,* and it is of three kinds. Thermal remanent magnetism occurs when rocks cool down through the Curie temperature; this type is found in basalts. Chemical remanent magnetism occurs when crystals precipitate from solution and is found in iron-rich limestones. Many sedimentary rocks acquire chemical remanent magnetism during diagenesis when the iron oxide mineral hematite (Fe_2O_3) precipitates between clastic grains. Depositional remanent magnetism occurs when magnetic crystals, settling from a fluid, orient themselves parallel to the magnetic field; this type is found in sedimentary rocks containing detrital iron oxides such as magnetite or hematite.

How can one detect remanent magnetism? By testing whether or not the rock behaves as a magnet. You can do this yourself. Take a sewing needle and rub it along a magnet until it is well magnetized. Then tie it, at its balance point, with a piece of thread, leaving six inches or more of thread on each end of the knot. Suspend the thread in some draught-free place so that the needle is free to rotate on the vertical thread, and let the needle settle down parallel to the Earth's magnetic field. (Unless you live on the equator, because of the magnetic inclination, the needle will not be horizontal.) Then bring a piece of basalt near the needle, which may be deflected by the magnetic field of the rock.*

The strength of the magnetic field created by a rock sample is seldom strong enough to deflect a magnetized needle. Magnetic measurements on ordinary rock samples are performed with sensitive instruments in specially constructed "field-free" rooms from which Earth's ambient magnetic field has been excluded. The measurement relies upon electromagnetic induction. As Michael Faraday (1791–1867) discovered, when a conductor crosses magnetic lines of force, an electromotive force develops within it, and if the conductor is part of a closed circuit an electric current develops. The strength of the current depends upon the strength of the magnetic field and the velocity at which the conductor crosses the lines of force. By turning a rock sample (the magnet) within a coil of copper wire (the conductor) at a specific velocity, one can determine the strength of the magnetic field the rock generates. By mounting the rock sample in different orientations, one can determine whether or not the rock magnetism is preferentially aligned.

If one collects oriented samples of appropriate rocks (e.g., basalts or iron-rich sandstones) of different, well-defined ages, one discovers that the Earth's magnetic axis seems to have wandered around a bit over the years. Worse yet, sequences of rocks from different continents indicate different paths of polar wandering for the same periods of earth history. Before the recognition of seafloor spreading and the Theory of Plate Tectonics, this was a baffling discrepancy, but after correcting for the displacements of the continents, themselves, the polar wandering curves for different continents fell into alignment. The magnetic poles do wander but never more than 20 degrees from the Earth's axis of rotation.

In 1906 French physicist Bernard Brunhes made a puzzling discovery. Remnant magnetism in certain basaltic lavas of the Auvergne—where D'Aubuisson, Von Buch and other neptunists had

*Basalt is a common igneous rock in most parts of the world, but if you live where the ground is soil or sedimentary or metamorphic rocks you might find a piece of basalt along a railroad track for which basalt is the preferred foundation material. A common name for basalt in such situations is traprock.

confronted the igneous nature of basalt—indicated that the north magnetic pole had been at the south geographic pole when the lavas crystallized. This was the first clear evidence of magnetic field reversal, but it was too far ahead of its time. The fact that the Earth's magnetic field reverses was not recognized for another half century. As Winston Churchill is reported to have remarked, "Men occasionally stumble over the truth, but most of them pick themselves up and hurry off as if nothing happened."[74]

The truth of the matter was recognized during the 1950s, after several other observations of reversed polarity. Every few hundred thousand to a few million years, the Earth's field reverses itself. The north magnetic pole moves to the southern hemisphere, and then, after another few hundred thousand to a few million years, it flips back to the northern hemisphere. So far as we know, this reversing of polarity has gone on for most, if not all, of geologic time. It provides time-signals of great accuracy in rocks all over the Earth and is extremely useful for correlating geological and archæological records from one place to another.

The most spectacular record of magnetic reversals is in the oceanic crust. Stripes of normal or reversed magnetic polarity parallel the accreting plate margins (mid-ocean ridges). Unfortunately, the record is only as old as the oldest oceanic crust, which is less than 200 million years. Nevertheless, for the last 200 million years we have a detailed magnetic record of the growth of the oceanic crust, and it is this record that has provided convincing confirmation of a new idea, *seafloor spreading*, and has thereby resurrected the notion of continental drift.

Eight

How Continents Move

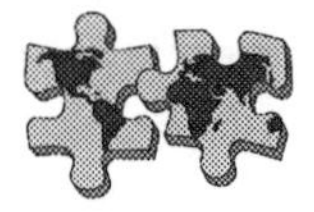

The history of the idea of continental drift and its relationship to the Theory of Plate Tectonics is an example of science progressing from observation to idea to test to failure and back to more observations, ideas, and tests. In 1915, German meteorologist and earth scientist Alfred Wegener (1880–1930) published a book called *Die Entstehung der Kontinente und Ozeane,* in which he suggested that the continents had moved around on the face of the Earth. In 1924 the third edition was translated into English and appeared as *The Origin of Continents and Oceans*. This was no mere pipe dream, although many scientists treated it as such. Wegener presented arguments from geodesy, geophysics, geology, paleontology, biology, and paleoclimatology. He noticed, as had several others before him including Sir Francis Bacon (1561–1626) in 1620, that the east and west coast lines of the Atlantic are remarkably parallel. According to his own recollection, however, he discounted the thought of physical connection as being too improbable. But then he learned something else:

> In the fall of 1911, I came accidently upon a synoptic report in which I learned for the first time of paleontological evidence for a formed land bridge between Brazil and Africa. As a result I undertook a cursory examination of relevant research in the fields of geology and paleontology, and this proved immediately such weighty corroboration

> that a conviction of the fundamental soundness of the idea took root in my mind.[75]

Fossils, especially in Africa and South America, suggested the widespread dispersion of creatures that could not have crossed wide oceans. For some scientists, these were compelling arguments for continental displacements; for others, they were not. One of the most knowledgeable and thoughtful paleontologists* of the time, George Gaylord Simpson (1902–1984), analyzed the fossil evidence and found it wanting. He offered various alternative explanations for the data and was severely critical of the more careless correlations. Of this period, Simpson later wrote:

> "I soon found (and this is still correct) that most of Wegener's supposed paleontological and biological evidence was equivocal and that some of it was simply wrong. (Wegener had no firsthand acquaintance with these fields.) Other early supporters of continental drift, all nonpaleontologists, even more grossly misrepresented the paleontological evidence. . . . in the 1930s and 1940s after lengthy investigation I concluded that the then-available real evidence of known land mammals not only did not support but opposed any effects of continental drift during the Cenozoic. . . . I did not deny the possibility of earlier effects of drift, but at that time I considered evidence for the drift theory so scanty and equivocal as to make it an unconfirmed hypothesis."[76]

Later, in the 1960s, Simpson changed his mind and accepted the idea of continental movements, but geophysical evidence, not fossils, had persuaded him.

Correlations of continental geology provided supportive evidence, especially in the southern hemisphere, but most geologists lived in Europe and North America and had not visited southern Africa or South America. Fortunately, however, some of the most capable geologists of the time were South Africans, and among these Alexander DuToit became a powerful advocate for continental

*Paleontology is the study of fossils.

drift. DuToit and others documented the striking similarity of Carboniferous and Jurassic sedimentary rock sequences in South Africa and South America. Progressing from glacial deposits to shales to coal beds to sandstones and finally basaltic lava flows, these sequences were also found in India and Antarctica. DuToit also showed that some structural trends, such as certain fold belts, which are presently interrupted by ocean basins, would be aligned and continuous if the intervening ocean basins were eliminated. A familiar example is the Appalachian-Caledonide belt that disappears beneath the North Atlantic off Newfoundland and reappears in northern Ireland and Scotland.

The evidence from marine geology was less persuasive because much less was known of the ocean basins. Wegener knew the general distribution of elevations on the planet and consequently knew that the Earth's topography is bimodal—with roughly one-third of the surface is close to sea level and two-thirds lie well below. He also knew about the Mid-Atlantic Ridge, but he did not know about its worldwide extent. And he knew about deep-sea trenches, but he knew nothing about the gravity fields associated with them. He knew little about the smaller-scale features of the seafloor, such as fracture zones, rift valleys, and submarine volcanoes. He knew nothing of the age or history of the oceanic crust.

The climatological evidence involved the occurrence of ancient glacial sediments at low latitudes and coal beds at high latitudes. This contradiction of the normal arrangement argued strongly for a different placement of the continents. Furthermore, the distribution of Paleozoic glacial deposits and structures in Africa, South America, India, and Australia suggested an ice source in the southern ocean, yet a continental glacier could not originate within an open ocean because the ocean currents would deliver too much heat to the source region. One solution was to move the continents.

Wegener also presented various geophysical and geodetic arguments in support of his idea. His geophysical arguments included the recently discovered phenomenon of *isostasy*, in which the continents appeared to be floating. If they could bob up and down to maintain isostatic equilibrium, surely they could also drift sideways.

Wegener's geodetic arguments attempted to demonstrate that specific locations on Earth had shifted in longitude or latitude within the brief history of astronomical records. His strategy was right, but his astronomical and geochronological data were wrong. His calculations were off by two orders of magnitude, suggesting displacements of meters rather than centimeters per year. Only within the last few years, in fact, have we had sufficiently precise methods to refer short-term plate motions to celestial reference points.

Wegener's story illustrates two important points about the practice of science: the fruitfulness of interdisciplinary study and the importance of doing what needs to be done. Disciplinary boundaries are human constructs, and Nature has little respect for them. Very often crucial observations or concepts are in the gap or overlap between two disciplines. As one wag said, "The grass may not be greener on the other side of the fence, but it is likely to be longer and more lush beneath the fence." The second point has to do with attitude, which is often the difference between a very good scientist and a not so good one. In the words of American scientist Arthur Holly Compton, "the difference . . . is whether he will make himself do the thing he sees needs to be done".[77] Wegener did not shy away from paleontology just because he was by training a meteorologist. He saw the need to learn paleontology, geology, and geophysics, and so he did. To do so took considerable courage, for Wegener must have known that he would make errors, especially in fields where he lacked firsthand experience. He did make errors, and his errors were revealed and ridiculed, but he also made and provoked progress.

According to the idea of *Continental Drift*, continents move, like great ships, through denser and weaker rocks that lie beneath the oceans. This idea was seriously flawed, because, as many scientists realized at the time, the cold rocks of the oceanic crust are too strong to allow the continents simply to plow through them. One of the most influential critics of continental drift was Sir Harold Jeffreys. He like Lord Kelvin, who played a similar role in the debate over the age of the Earth, is sometimes cast as the heavy, the one who obstructed science, but this is both wrong and unfair. The Kelvins and Jeffreys of science, who insist upon rigorous testing of ideas, protect science from dreams and dogmas. Of the debate itself, Jeffreys wrote:

> It is argued that if evidence from palæontology and meteorology proves that continental drift has taken place, evidence from geophysics is that it is impossible is beside the point. If I admitted the premises I might accept the conclu-

> sion, while still maintaining that it is remarkable that the advocates of continental drift have not produced in thirty years an explanation that will bear inspection. But I must reject the whole attitude that maintains that any type of scientific evidence can by itself be so completely demonstrative as to require the rejection of any evidence that appears to conflict with it. If evidence is conflicting, the scientific attitude is to look for a new idea that may reconcile it. The question here is whether the new idea is to be found in palæontology, meteorology, or in the mechanics of the Earth itself. I think myself that it is least likely to be in the last, because it is the easiest subject of the three. [78]

And, ironically, that's exactly where the new idea was found, in the mechanics of the Earth itself!

The new idea, the Theory of Plate Tectonics, resulted from observations made over the fifty years following Wegener's initial contribution. Prior to World War II, earth scientists, including Alfred Wegener, were severely limited by their ignorance of the deep ocean basins. They had only vague clues about their shape and the nature of their bedrock, and they had virtually no knowledge of their history. Sir Edward Bullard (1907–1980), who turned his attention to the ocean basins in the 1930s, later recalled the situation:

> On the whole Wegener's opponents had the best of the pre-war arguments. During the 1930s and 1940s it was unusual and a little reprehensible to believe in continental drift. It is easy now to see that what was needed was not further disputes about the old arguments, which had been demonstrated not to carry conviction, but new evidence. . . . Soon a great discovery was made; or rather gradually became apparent; the oceans are quite different from the continents."[79]

In retrospect, we see how fields as disparate as micropaleontology, isotope geochemistry, and metamorphic petrology contributed to the new idea, but, as it developed, the pace and direction of the conceptual revolution were set by discoveries in geophysics and marine geology.

In geophysics, the principal discoveries were in paleomagnetics, heat flow measurements, and seismology. Paleomagnetic studies on the continents indicated that the magnetic pole had shifted regularly from one location to another throughout the geologic past; however, different continents showed *different* paths of "polar wandering"! Heat flow measurements indicated that, on average, as much heat was flowing from the oceanic crust as from the continents. This was surprising given the higher concentrations of radioactive atoms in continental crust and the thinness of oceanic crust. It was also observed that by far the greatest oceanic heat flow was coming through the mid-ocean ridge crests.

As the number and quality of seismic data increased it became apparent that most earthquakes occur along narrow lanes, extending fewer than 100 kilometers beneath the Earth's surface, or along inclined planar zones that descend to depths of 600 to 700 kilometers beneath island arcs, the west coast of Central and South America, and parts of the southwest Pacific. The narrow lanes of shallow quakes were along the mid-ocean ridges, which were seen to girdle the Earth like the seams on a baseball, and along great transcurrent faults, such as California's San Andreas Fault and New Zealand's Alpine Fault. The inclined zones of earthquakes were associated with trenches in the sea floor. Even before World War II these inclined zones of earthquake activity had been identified as possible locations of very large thrust faults.

The invention of the echo sounder provided the means to map the shape of the sea floor in detail. In 1958 Bruce Heezen and Marie Tharpe, marine geologists at the Lamont Geological Observatory near New York City, observed a narrow valley running down the center of the Mid Atlantic Ridge. Heezen compared profiles of this submarine valley with the Red Sea and the great East African rift valleys and suggested that the Atlantic Ocean was splitting apart at the center. His suggestion, offered at a seminar at Princeton University, was met with considerable skepticism because it seemed to require expansion of the planet. One of the students in the audience recalls the scene:

> [He was sitting] in the back of the room thinking this was all nonsense. Heezen's axial valley looked dubious, the expanding earth was physically impossible, and so why take the East Africa to Red Sea to Atlantic Ocean progression seriously? In fact there was some curiosity about what Harry Hess, who was the chairman that evening, could say in the way of being polite after a lecture that was so impossible to accept. At the end of Heezen's lecture, Hess stood up and said, "Thank you for a lecture that shakes geology to its very foundations." Some in the audience thought Hess had merely found a clever way of being polite. But events soon showed that Hess meant his compliment. [80]

Harry Hess (1906–1969) was a geologist of wide experience and deep insight. As a young man he had accompanied Dutch geodesist and geophysicist F. A. Vening Meinesz into the Puerto Rican trench in a submarine in order to measure gravity (little *g*). They found, as Vening Meinesz had found in oceans all over the world, that the trench axis corresponded to a pronounced reduction in gravitational attraction—a so-called *negative anomaly.* Reduced gravity ought to encourage matter to rise up, away from the center of attraction, yet here was a trench. The surface of the Earth had somehow been drawn down, closer to the center of gravity!

Hess returned to the sea for two years during World War II as Commander of USS Cape Johnson. While his ship steamed back and forth across the western Pacific he kept the echo sounder turned on, so that he and his crew collected miles and miles of deep-sea soundings. In 1944, he observed the first of "a large number of flat topped peaks . . . roughly oval in plan . . . truncated by a level surface which now stands approximately 750 fathoms (4500 feet) below sea level."[81] These peaks he named *guyots,* after a nineteenth century geographer at Princeton. At first, Hess thought guyots were very old volcanoes, perhaps even of Pre-Cambrian age, which had been eroded to sea level and subsequently drowned as accumulated sediments raised the level of the sea surface. However, 20 years later he saw them as evidence that the sea floor had moved laterally away from the shallow mid-ocean ridges and sunk. The guyots and the deep sea trenches lay in Hess's mind until 1960 when he put the puzzle of continental drift and oceanic crust evolution together in a paper he referred to as "an essay in geopoetry."

In this paper, *History of the Ocean Basins*, which was published in 1962, Hess suggested that the deep ocean trenches, such as the Puerto Rican trench, mark zones of crustal subduction where the oceanic crust, created at the mid-ocean ridge, is drawn back into the planet and destroyed. He thereby eliminated the need to suppose any expansion of the planet. As crust was created, it was consumed. The continents did not plow through the oceanic crust, as Wegener had supposed, but were carried along passively on the top of large convection cells. Hess shared his thoughts with Robert Dietz, who further developed the idea and invented the term *seafloor spreading*. The concept of seafloor spreading developed rapidly following Hess' and Dietz' papers.

Harry Hess was one of the dozen or so architects of the Theory of Plate Tectonics, beginning with Alfred Wegener. To those who knew him and some of the others, however, Hess stands apart. He was a special person, who inspired devotion among his students and colleagues such as Abraham Gottlob Werner had done two centuries earlier. He was a person endowed with extraordinary intelligence, which was refined by education and hidden by shyness, a passionate scholar who was able to concentrate longer and more deeply than others, and a gentle man blessed with heartfelt concern for others. The accounts of Werner suggest that he shared some of these fine attributes, but there was a fundamental difference between the two men: Werner never acknowledged his errors, and Hess always did. After careful consideration, over many years, Hess could (and more than once did) turn away from a cherished idea and move on towards some less refined but more viable concept.

Hess was taciturn. He could sit *forever* without saying a word. I got used to this and learned to keep my own mouth shut, but others found his silences disconcerting. Once on a morning drive to a graduate student's field area Harry was deep in thought. While the student peppered him with questions, he said nothing and the student continued to talk. Later, as they sat on a hillside, the student became even more confident and expansive about his ideas. The hill that morning was covered with butterflies. At last Hess spoke, "You know, if you aren't careful, you might swallow one of those." Neither of them said much more for the rest of the day. The next morning they set off again. The student was subdued; Hess was smoking quietly and enjoying the scenery. As they bumped along in their Jeep, he began to speak, answering each of the student's questions from the day before, comprehensively and in sequence.

More disconcerting than silence was Harry's tendency to omit sentences. He would complete a thought in his mind but neglect to express every essential point out loud. Another brilliant scientist, Jason Morgan, the co-inventor of plate tectonics, who had joined Harry's department at Princeton, told me after Harry's death that he had not always understood what Hess said. "Oh, Jason," I exclaimed, "you didn't know? Harry generally left out a sentence or two!" Jason was visibly relieved.

Harry didn't write much either, and he took his time to publish. His great "essay in geopoetry" that introduced the idea of seafloor spreading languished for two years in a *festschrift* for his friend and mentor Arthur Buddington as the ponderous wheels of monograph publication ground slow but fine. It was only thanks to the decency of Robert Dietz, whose report of the idea preceded Hess' publication, that Hess received uncontested credit for the concept. On the surface Hess seemed not to care. I once asked him if he would coauthor a report of work we had done together. "No," he said, "and when you become a professor you mustn't do that either."

Hess' office at Princeton was a mare's nest of charts, reports, and notebooks. He never discarded anything. On the wall were mounted the horns of a gazelle. He had brought them home from Africa years before, uncured and smelling badly. Over the years they had dried, ceased to smell, and took their place with bits and pieces of long-finished researches lying about the room, graphs and charts mostly, not many rocks. Harry warned me against collecting reprints. "It's a terrible habit," he said.

Conversations with Harry could be disjointed, as one of his students describes: "More than once he approached me with such opening gambits as 'No, I don't think so.' I was invariably taken aback and only after several exchanges would it emerge that he was referring to casual opinions made many months previously in the field in Venezuela. It is said he once picked up in mid-stream a discussion interrupted only by World War II!"[82]

Hess was deeply patriotic and proud of his military service and status. He had commanded a ship in World War II and was a Rear Admiral in the Naval Reserve. One night, under the stars at a mining camp in the hills of Venezuela, we talked about the festering war in Vietnam. Like many university students of my generation, I was opposed to the American effort; Harry disagreed. As we parted for bed I said that I had more confidence in his judgment about rocks than about Vietnam. "Well," he smiled, "I know a lot more about rocks."

Harry loved to smoke Chesterfields and he liked to drink. We used to measure hikes in cigarettes. "How far is it?" "Oh, that's about three cigarettes, Harry." "OK, let's go." So we'd climb the trail and stop three times for a smoke and a chat. The cigarettes and alcohol shortened his life. That, I am sure, was his conscious choice. After his first heart attack his doctor told him to cut down on his smoking and to avoid whisky. Out of deference to his physician, he switched to brandy. Although he neglected his own well-being, Harry Hess cared for many people, especially his students. He understood their needs, strengths, and weaknesses with great sensitivity. I don't believe anyone, except possibly his wife, understood him as well as he understood others. He was too complex, too deep, and too shy for even a close friend to completely comprehend.

For reasons that remain unclear, heat flowing from within the planet concentrates along narrow channels, causing the surface to rise in mounds or ridges. Eventually gravity induces the cold outer skin of the planet to slide downhill away from the raised region. Because rock has little tensile strength, especially when hot, the cold skin cracks and pulls apart above the rising heat, and immediately hot material rises to fill the void. Decompression favors melting in hot dry rocks, so the rising material, which may have been partially molten to begin with, continues to melt as it rises. When the white hot slush of residual solid matter and molten rock rises to a depth of 10 to 20 kilometers, the liquid, now constituting perhaps 15 percent of the material, segregates from the slush and streams rapidly upwards, some to flow out on the surface as lava, some to crystallize below the surface as intrusive igneous rocks. The residual (unmelted) solids, stripped of their low-melting constituents, are eventually pushed to the side by the next pulse of rising mantle material.

The process just described may begin beneath a continent, as in East Africa, but because the rising, melting mantle displaces less dense continental rocks, the new-formed surface lies low on the planet, below sea level, and eventually becomes an ocean basin.

Early in the twentieth century a German geologist, Gerhard Steinmann, noticed a recurring association in mountain belts of radiolarian cherts, metamorphosed pillow basalts, and serpentinized peridotites. Without realizing it, he was identifying the products of sea floor spreading and their associated deep sea sediments. Harry Hess spent much of his life studying these rocks, known as "Steinmann's Trinity." That, and his oceanographic experiences, prepared him to conceive of seafloor spreading. As Louis Pasteur remarked, "Where observation is concerned, chance favors the prepared mind."*

Just at the time Hess and Dietz were publishing their ideas about seafloor spreading, three geophysicists were studying a strange striped pattern of magnetic signals on the ocean bottom. Fred Vine, a graduate student, and his advisor, Drummond Mathews, were at Cambridge University in England; Lawrence Morley was an aeromagnetics expert at the Geological Survey of Canada. Vine and Mathews were working with data they had collected over the Rekjanes Ridge south of Iceland. Morley was studying data collected by others in the east Pacific. Morley tells the tale:

> By 1961 . . . it was becoming obvious that magnetic banding in the ocean basin was the rule rather than the exception. For the next few weeks I could think of nothing else. My regular duties were totally neglected as I searched the literature for clues. It was obvious that there was some explanation that was fundamental to the whole origin and

---

*Dans le champs de l'observation le hazard ne favorise que les espirits préparés."[83]

geological structure of ocean basins. I became obsessed. Why was there a regular banding pattern? Why was it so different from the convoluted patterns over the continents, with which I was so familiar? . . . From my knowledge of rock magnetism . . . I knew that this positive and negative banding had to be due to remanence. . . . These data sat in the literature for at least 3 or 4 years with no explanation . . . I was sure that the cause was remanence and that the positive and negative banding was associated with the possible periodic reversals of the earth's field . . . Still there was no complete explanation. In searching the geological literature on the ocean basins, it struck me that this banding might somehow be related to the East Pacific Rise, simply because the banding was parallel to the ridge. Then I ran across Robert Dietz's paper on ocean floor spreading (Dietz, 1961). Eureka! I knew immediately that this was the explanation. If the rocks at the midocean ridges were rising from depth, they would become thermoremanently magnetized in the direction of the earth's field prevailing at the time. They would then spread laterally in both directions towards the continents . . . A million or so years later the earth's field would reverse, and this way a positive and negative banding pattern would be built up. . . . Over the next 8 months I tried desperately and unsuccessfully to get my idea into print. I first submitted my paper to Nature and it was rejected on the grounds that the journal did not have enough room! I submitted it to the Journal of Geophysical Research . . . I received a rejection notice from the editor accompanied by an enclosed note . . . The note apologized for the long delay . . . [The reviewer] said "it was an interesting idea . . . but was something which was more appropriately discussed at a cocktail party than published in a serious scientific journal." Just as I was planning to submit it to a Canadian journal, the September 7, 1963 issue of Nature came out with an article by Vine and Mathews (1963) entitled "Magnetic Anomalies over Oceanic Ridges," giving the same explanation that I had. For me, the main "ball game" was over.[84]

The race to publish has always been, and will always be, inherent to scientific research. In this respect science and art are very different. Because scientific achievements must be reproducible, there can be no irreplaceable treasures in science. Priority is important, but to be overly concerned with it is to flirt with hubris. In the long run Morley, Vine, and Mathews will all be recognized, remembered, and forgotten. What will not be forgotten is that the magnetic patterns on the sea floor confirmed the concept of sea floor spreading, which was a major step towards the Theory of Plate Tectonics.

As the story (which may be apocryphal) goes, Jason Morgan, a young geophysicist on the Princeton faculty, wandered into a graduate student's office one morning and said, "I've just noticed something interesting. These fracture zones across the Mid-Atlantic Ridge lie on small circles, and they appear to have a common pole. Would you like to help me work this out?" The student declined—he had his comprehensive exams coming up. He passed his exams, and Morgan went on to formulate the Theory of Plate Tectonics, sharing this achievement with French geophysicist Xavier LePichon, who simultaneously, with a slightly different approach, developed the same ideas. Morgan and LePichon demonstrated that terrestrial tectonic phenomena are interrelated on a global scale because the lithosphere behaves as a strong, rigid shell, broken into a small number of pieces, called plates, that are constantly renewed, displaced, and consumed. The Theory of Plate Tectonics unified a vast and varied body of data and offered extraordinary opportunities for prediction and test.

Unlike James Hutton's Theory of the Earth, or the Theory of Evolution, which was simultaneously and independently created by Charles Darwin and Alfred Wallace, the Theory of Plate Tectonics was conceived almost simultaneously by several scientists. Morgan and LePichon are usually given credit for putting it all together in 1968, but J. Tuzo Wilson had divided Earth's surface into "several large rigid plates" in 1965, and in 1967 Dan P. MacKenzie and R. L. Parker had published ideas quite similar to those of Morgan and LePichon. That's the way science works nowadays. The search for truth is seldom lonely, and when the hunt closes in it's a merry mob scene.

The lithosphere, the outermost rock shell of Earth, is cold enough and therefore strong enough—rocks get weak when they are hot—to absorb elastic strain energy and to behave for significant periods of time as a rigid solid. It is approximately 100 kilometers thick in oceanic areas and considerably thicker in many continental regions. The lithosphere includes both the crust and the upper mantle, its base is generally identified with the low velocity zone, where melting and plastic deformation of rocks are thought to be common.

Mechanically, the Theory of Plate Tectonics differs fundamentally from the Theory of Continental Drift. Drifting continents were to move like rafts in oceanic crust, which is physically impossible. According to the Theory of Plate Tectonics, the moving slabs are more than twice as thick as the crust and therefore much stronger. Plate theory also uses a different mechanism for displacement. The plates do not push their way through cold, strong crust; when they eventually do push material aside, it is hot, weak mantle.

The lithosphere is subdivided into a dozen or so large, horizontal pieces that are shaped like pieces of a broken bowl. These plates are outlined by zones of frequent earthquakes and volcanic activity. What Morgan and LePichon (and MacKenzie and Parker) proved was that these plates function as rigid units, moving around on the surface of the Earth without undergoing much internal deformation. Most of the action is confined to the plate margins. There are three kinds of plate margins: accreting, consuming, and shear, which turn out to be transform faults.

New material is added to plates in zones along the mid-ocean ridges. These zones are called *accreting margins* because matter is accreted onto the plates. Accreting margins experience frequent shallow earthquakes associated with the movement of the lithospheric plates away from the mid-ocean ridges. Normal faulting is common in this pull-apart environment; the plates move a few centimeters per year, about as fast as fingernails grow.

As the plates move away from the hot, relatively shallow mid-ocean ridges, they cool down, and as they cool they become more dense. As the lithosphere becomes denser, it sinks towards the Earth's center of gravity, which is why the ocean basins deepen regularly with distance (in space and time) from the ridge. This relationship, predicted by geophysicists Jean Francheteau and John Sclater in 1970 and subsequently confirmed, is a good example of the kind of peripheral explanation that derives from a great unifying theory. Eventually the plates sink back into the body of the Earth. The line along which this subduction occurs is marked by the great oceanic trenches with their overlying volcanic island arcs. Earthquakes accompany the sinking slabs to depths as great as 680 kilometers. How far the slabs sink we do not know for sure. Some seismologists believe they see evidence of slabs sinking right down to the mantle/core boundary, 2,900 kilometers below the surface. Others suggest that they descend no deeper than 680 kilometers. The consuming plate margins are where two lithospheric plates move towards each other. For that reason they are under lateral compression. The topside of the downgoing slab may be thought of as a colossal reverse fault.

Lithospheric plates may slide horizontally past one another along great transverse faults. The San Andreas fault of California is our most familiar example. Hidden beneath the sea are thousands of smaller, transform faults, which chop the mid-ocean ridges into short segments. These transform faults allow the lithospheric plates to compensate for short term variations in spreading rate along the accreting margins. There are also scars on the ocean floor left by transform faults. These scars are not active faults—only fault segments that connect active spreading centers are active—but, as Jason Morgan and others showed, they record the directions of plate motions in the past. The enormous Pacific plate, which extends from California to Japan, includes thirteen transform fault scars that score the oceanic crust along east-west arcs, mimicking lines of latitude, from the Gulf of Alaska at 45 degrees north to the latitude of New Zealand at 45 degrees south.

Among the terrestrial planets, the Earth is distinguished at present by prolific life, abundant surface water, a dynamic interior, and linear mountain belts. Mountains on the Moon and Mercury form rings (circles), not belts. Mountains on Mars do not form belts, either, but great mounds and plateaus. Only Venus has mountain belts that at all resemble those of Earth. Mountain ranges occur as belts on Earth because they form along great linear features—the accreting and consuming plate margins.

In terms of length, volume, or mass, the Earth's mid-ocean ridge (MOR) mountain chain is the largest mountain belt on the terrestrial planets. It is exposed to us only in places where exceptional mountain tops rise above the sea—Iceland is the most extensive example. The structure of the MOR mountain system is relatively simple. Steep faults are common and folds are absent. The MOR mountain system exists because great quantities of Earth's internal heat rise beneath the spreading centers, and that heat causes the surface of the planet to bulge outwards in a long, low-density welt.

Arcuate strings of volcanoes, such as the Aleutians, constitute another kind of mountain belt on Earth. These are located over the magma-producing regions associated with subduction.

Belts of folded mountains, such as the Alpine chain extending from the Pyrenees to Sumatra, are manifestations of incomplete subduction. All sedimentary and some igneous and metamorphic rocks are too buoyant (that is, not dense enough) to sink very far into the mantle with the downgoing lithospheric plate, so there is a strong tendency for them to accumulate in piles above subduction zones. Eventually such piles are squeezed into complex melanges of material, which are pushed up and mixed in with rocks of the overlying (and overriding) island arc or continent. The Andes of South America, the New Zealand Alps, and the mountains of Taiwan are just a few examples of mountain belts created by such processes.

When two plates that are both carrying continents approach one another over a subduction zone, a major collision of low density materials occurs. The result is a major folded mountain belt such as the Appalachians, the European Alps, or the Himalayas. The Himalayas, for example, were created by a collision between the continental components of the Eurasian and Indian plates.

Of the many scientists who participated in the development of the Theory of Plate Tectonics, none was more creative or influential than J. Tuzo Wilson of Canada. Wilson's career was energetic and brilliant throughout, but his most important contributions came after he was persuaded by Harry Hess's 1962 paper that lateral mobility was not outlandish nonsense but the great reality of terrestrial tectonics. Wilson was then fifty-four years old. Although he is best known for the concept of the transform fault, his overarching contribution was observing that Earth's great linear features—mountain belts, oceanic trenches, and great faults—are integral, interconnected manifestations of a global tectonic process. This was the intellectual insight that led directly to the Theory of Plate Tectonics.

During the exciting years immediately after the theory had been proposed, when all sorts of unexpected applications were discovered and exploited, Tuzo Wilson, with his younger colleague Kevin Burke, developed the concept of a cycle connecting the processes that form ocean basins and mountain belts, known as the *Wilson Cycle*.

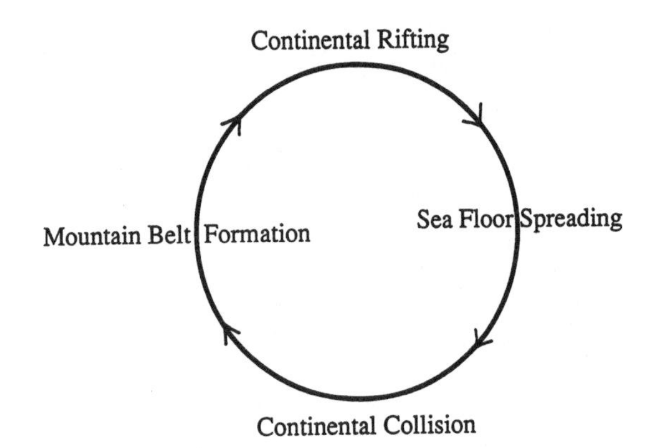

There have been three major theories influencing our understanding of the Earth: Hutton's Theory of the Earth, Darwin's Theory of Evolution, and the Theory of Plate Tectonics. Each provided sudden, almost breathtaking connections between what had previously been disparate and confusing observations. Each opened entire new fields for experiment, observation, and analysis.

Nine

Ores, Energy, Radwaste, and Water

Earth supplies all of our material commodities. Modern societies use most of the 92 naturally occurring elements, and new combinations are replacing traditional forms in many applications. One way or another, however, all materials are derived from Earth's resources.

Ores are concentrations of mineral resources that can be mined and sold for a profit. They occur in igneous, metamorphic, and sedimentary rocks. Examples are the igneous ores of chromium, the metamorphic ores of copper, and the sedimentary ores of gold.

An average sample of Earth's crust contains less than 0.01 percent chromium. There is more chromium in the mantle; we don't know for sure, but average mantle probably contains more than 0.3 percent. Chromium ore deposits form when the mantle melts and forms basaltic magma, which subsequently precipitates the mineral chromite. Chromite, a solid solution of $FeCr_2O_4$, $MgCr_2O_4$, and $MgAl_2O_4$, is 50 percent more dense than the basaltic magma, so it sinks rapidly and forms sediment-like deposits on the floors of magma chambers and conduits. Occasionally, these deposits are exposed on the Earth's surface by uplift and erosion. The largest chromite deposit on Earth is in the Transvaal region of the Republic of South Africa, where a huge, 2-billion-year-old magma chamber, known as the Bushveld Igneous Complex, has been exposed over an area of 65,000 square kilometers. Within this eroded pile of igneous sediments are more than 6 billion tons of ore containing more than 40 percent chromium oxide.

Many ore deposits, especially those of base metals (e.g., copper, lead, and zinc), are created when hot water moves through vast volumes of rock, dissolves the metals, and reprecipitates them in greater concentrations elsewhere. Such processes are called *hydrothermal.* Among the largest and most interesting hydrothermal ore deposits are those associated with porphyritic intrusions of siliceous magmas. The term *porphyritic* refers to an igneous texture in which large crystals lie in a groundmass of much smaller crystals. This texture is interpreted to mean that the large crystals grew in the magma, for some extended period, before it erupted, or moved to a cooler intrusive environment, and quickly froze. Earth's largest known copper deposits are of this type, referred to as *porphyry copper deposits.*

Porphyry copper deposits are associated with the siliceous magmas that feed andesitic or rhyolitic volcanoes over subduction zones. The American Cordillera, from Alaska to Chile, includes more than 75 major such deposits. The concentrations of copper tend to be low (less than 1 percent), but the quantities are huge. The largest deposit, at Bingham, Utah, contains more than a billion tons of ore.

A generalized sequence of events for forming a porphyry-type deposit is as follows: Wet, siliceous magma rises into the crust above a subduction zone and begins to crystallize. A volcanic eruption causes the magma to rise rapidly in the conduit below the volcano, and the sudden release of pressure causes the wet magma to freeze, forming a cyindrical intrusion called a *stock.* Superheated steam explodes from the quenched (rapidly cooled) magma, thoroughly fracturing the hot rocks. Driven by heat from the intrusion, water convects around and through the stock. Base metals, especially copper, are leached from the surrounding crustal rocks and from the intrusion itself. Eventually these metals are reprecipitated in zones resembling inverted tea cups over and around the top of the stock. Depending upon the temperature and pressure at various locations in the convection system, the aqueous fluid may be liquid or gas. Decompression of rising hot water causes boiling, so higher concentrations of ore minerals tend to occur where the fluid boils because a gas can carry much less dissolved ore material than can a liquid.

In addition to copper, porphyry-type deposits contain many other metals, such as molybdenum, lead, zinc, silver, and gold.

Gold is brought to the crust in magmas and concentrated by hydrothermal processes in veins, commonly with quartz. Eventually these veins are exposed to weathering and erosion, and grains of gold are released and carried towards the sea along with other resistant (relatively insoluble) detritus. The gold is much more dense than quartz or other detrital silicates (19 g/cm^3 compared to 2.6 g/cm^3), so it settles through the gravel, sand, and mud and accumulates in depressions and crevices in the underlying bedrock.*

Accumulations of heavy sediments such as gold are called *placers*. Placer deposits have been mined for gold since the Neolithic Period, 8,000 to 4,000 years ago.

The Earth's largest known gold deposit is in thin conglomerate beds, or "reefs," that are part of an 8-kilometer-thick section of steeply dipping Precambrian sediments in the Witwatersrand region of the Republic of South Africa. Although there has been some recrystallization and remobilization of the gold, the deposit originated as placer sediments. Known as the Rand, it has produced more than half of the world's gold since 1886. The ore averages twelve grams of gold per ton (0.0012 percent).

The role of mining and mineral resources in the world is changing. In the past the wealth and power of nations, as well as the course of history, was directly affected by the natural distribution of material commodities. During the 1970s there was concern over "limits to growth" imposed on society by the finite material resources of the Earth.[85] However, that particular problem seems to have lessened for rich nations. In summarizing an analysis of the future demand for aluminum, chrome, cobalt, copper, iron, manganese, nickel, platinum, steel, tin, tungsten, and zinc, economist Wilfred Malenbaum writes as follows:

---

*Amateur gold panners would waste less time and effort if they would dig deep and pan the sand and gravel that lies directly on bedrock or, better yet, within bedrock cracks and crevices. Panning surficial sand and gravel is futile.

> The intensity of use (I-U)* of most materials and in world regions where most of each material is used has been declining over recent decades. A notable exception is aluminum: its I-U shows declining rates of increase. These patterns for the world are clearly manifest in rich nations. They are less definitive in the poor nations.
>
> The forces responsible for the declining pattern of intensity are:
>
> (1) Shifts in the types of final goods and services that world consumers and investors demand directly;
>
> (2) Technological developments that alter the efficiency with which raw materials are discovered, extracted, processed, distributed and utilized in production of final goods; and
>
> (3) Substitutions among raw materials inputs consequent upon relative price movements and relative rates of technological development.
>
> . . . The I-U evidence constitutes strong support for the argument that man's knowledge, skill and aspirations have served to slacken his need for industrial raw materials.[86]

In today's global economy, educated people, not mines or quarries, are the essential resource of a developed nation. Access to commodity markets has supplanted domestic resources as a necessary condition for commercial success. Indications are that, as developed nations become more advanced, their per capita consumption of minerals and metals decreases. Their per capita consumption of energy, however, increases.

The energy we liberate from fossil fuels came from solar energy stored in carbon, hydrogen, and oxygen compounds of living organisms and from terrestrial heat used to transform that plant and animal matter into coal, oil, and gas. The terrestrial heat

---

*Intensity of use is the amount of metal or mineral consumed in producing a unit of GNP.

source is frequently overlooked, but it is the source of a significant portion of the energy in fossil fuels. Take coal for example:

Comparative Heat Energy Content (cal/kg)[87]

Peat	5.5×10^6
Lignite	6.5×10^6
Bituminous coal	7.8×10^6
Anthracite coal	8.6×10^6

The increased heat energy content of anthracite compared to peat comes from Earth's internal heat. Fossil fuels are hydrocarbons produced by the elimination of water and oxygen and the reorganizing of the remaining elements into different (typically simpler) compounds. To make any of these fossil fuels, you need starting material rich in carbon and hydrogen, a depositional environment where there is little free oxygen, and mild conditions of prograde metamorphism (i.e., change brought about by increases in pressure or temperature).

Coal forms from freshwater plant remains. Environments favorable for the first stages of coal formation are swamps with stagnant (oxygen-poor) water. In order for the plant life to be abundant in the first place, there must be plenty of sunlight and water. Coal beds are typically overlain by clay, which protects the dead leaves and other vegetable matter from oxidation. As with most metamorphic processes, the first step in coal formation involves compaction and dehydration. As the deposit is more deeply buried, the temperature and pressure rise, driving off more and more water and reorganizing the coal-forming materials in a succession of compounds that contain less and less oxygen and hydrogen and more and more carbon. As this metamorphism proceeds, the carbon-to-oxygen ratio changes from approximately 1 to 1 in wood to 30 to 1 in anthracite. Ultimately, as the metamorphic process proceeds to higher temperatures and pressures, the organic matter will be converted to graphite, which is pure carbon.

Coal beds, being sediments, are typically thin and extensive. Consequently, the safest and cheapest way to exploit them is often by surface mining (stripping). The cost of mining is not as low as you might suppose, however, for it must also include the cost of returning the land to an environmentally acceptable condition.

The origin (or origins) of oil and gas are less certain than the origin of coal. Most geologists believe that oil and gas form from marine organisms such as phytoplankton, the plant-like, microscopic organisms that live in the sunlit waters of the oceans. Some geologists believe that some proportion of oil and gas is not biogenic but comes from reduced hydrocarbons, perhaps methane, that flow from Earth's deep interior. All evidence to date, from the optically active organic molecules in petroleum to the paucity of any organic substances in mantle samples, however, points to biogenic origins.

Like the raw materials for coal formation, the organic material for oil and gas formation must be deposited where there is little free oxygen. Restricted basins, rather than open ocean basins, may provide appropriate environments. Such environments also yield evaporite (salt) deposits, which frequently accompany oil and gas. Clay may act as a catalyst, promoting the oil- and gas-forming reactions, which occur as the sediment is buried. Shale is a common source rock for oil and gas.

As petroleum evolves through the chemical process, it changes from heavy crude, to light crude, to gas. The organic molecules become smaller. Common compounds in petroleum are paraffins (C_nH_{2n+2}) such as methane (CH_4), napthenes (C_nH_{2n}), and benzenes (C_nH_{2n-6}). Although petroleum must be created by gentle heating, it must not be heated too much. Overheating (greater than 150°C) destroys oil and then gas.

The oil and gas must be collected in a reservoir rock. Common rocks sufficiently porous to hold substantial quantities are sandstones and some limestones. Because oil and gas are less dense than water or rock, they tend to rise to the surface. In order to prevent this, there must be some kind of impermeable barrier, called a *trap*. Traps are usually formed by impermeable rocks such as salt or shale, which have been folded or faulted so as to create an enclosure above and alongside the hydrocarbons. Anticlinal folds (those that are shaped like inverted *U*s), for example, create excellent traps.

When certain heavy atomic nuclei are struck by neutrons with the right amount of kinetic energy, they split into roughly equal fragments and emit a few neutrons, which may themselves cause disintegration of other atomic nuclei. This is nuclear fission. Nuclear fission is accompanied by a tiny loss of mass, which is converted to energy according to the equation $E = mc^2$. That is the source of energy in nuclear power plants. The fuel used in nuclear power plants is uranium isotope 235 (^{235}U). A typical fission reaction is:

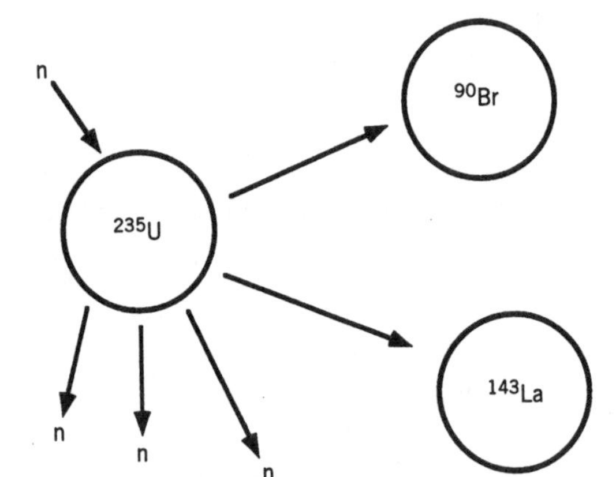

In a nuclear power plant, fission occurs within fuel rods of uranium dioxide, which contains ^{238}U enriched with a few percent of fissionable ^{235}U. The rate at which fission occurs depends upon the concentration of ^{235}U and the number and velocity of free neutrons. The neutron flux is controlled by inserting rods of cadmium or boron, which absorb neutrons. As the control rods are removed, more fission occurs and more power is released. Sufficient energy could be released to have a meltdown but not a nuclear explosion. Nuclear explosions require fuel with more than 95 percent ^{235}U.

Most nuclear power plants are less efficient than most fossil fuel power plants. This is because the fossil fuel plants can drive the turbines with steam direct from the boiler while most nuclear plants must drive the turbines with steam from a secondary circuit, which is heated by a primary (reactor-heated) circuit through a heat exchanger. Nuclear plants, therefore, may waste 40 to 50 percent more heat than comparable fossil fuel plants. On the other hand, the advantage of nuclear plants is that 1 gram of ^{235}U has the energy content of nearly 3 tons of coal or 14 barrels of oil.

The most serious disadvantage of nuclear energy is not the threat of a meltdown or the relatively low efficiency of the power plants. It is the production of radioactive waste, or *radwaste*. How to deal with radwaste has been one of the transcendent questions of science and technology policy since World War II. This issue is comparable in gravity and scope to the challenges of AIDS and the population explosion, comparable in technical complexity and audacity to placing human beings on the Moon, and comparable in political divisiveness to the issue of abortion.

Radwastes come as solids, liquids, or gases, and are characterized as low-, middle-, and high-level. Every ton of nuclear fuel produces about a ton of waste, of which 15 percent is high-level and long-lived. Different strategies apply to the different levels.

For low-level wastes, such as gas from stacks and coolant waters, the strategy is to dilute and disperse them. The dilution and dispersal procedures are based upon strict maximum permissable concentration limits (MPC), which are based upon fifty years' ingestion. The most serious problem here is biological reconcentration. Aquatic organisms may reconcentrate the substances by as much as 1,000 times. Low-level wastes in solid form (e.g., irradiated tools and furniture) cannot be easily diluted and dispersed and so are buried in drums for a long time.

For middle-level wastes, the strategy is to store them until they are safe to disperse.

High-level radwastes are a different kettle of fish. These substances are sufficiently radioactive to kill within minutes.* Some are

---

*There are several measures of radiation and degree of hazard. One curie is radioactive decay at a rate of 3.7×10^{10} disintegrations per second, which is approximately the radioactivity observed in one gram of radium. Maximum permissable concentrations (MPCs) are usually expressed as picocuries (10^{-12} curie) per liter. For ^{137}Cs (cesium) in air, the MPC is two picocuries per liter. Because it emits alpha particles, which are highly destructive to biological tissue, and because it is readily absorbed in bone marrow, ^{239}Pu (plutonium) is among the most dangerous constituents of radioactive waste. The MPC for ^{239}Pu in air, in terms of radioactivity, is 0.0006 picocuries per liter. High-level wastes produce hundreds to thousands of curies.

highly toxic chemical poisons as well; some are biologically active and easily absorbed; and some decay slowly and must therefore be contained for hundreds to thousands of years.

Although the quantities of high-level radwastes are small compared to many other toxic wastes, they are substantial. By the year 2000 the world's inventory of high-level radwastes from spent fuel is expected to exceed 280,000 tons, or 150,000 cubic yards. Of this, the United States will have more than 45,000 tons, or more than 24,000 cubic yards. In addition, the United States will have more than 80 million gallons of high-level waste in nitric acid solution from reprocessing plants.[88]

The disposal of radwastes poses unique challenges to our ingenuity. As has been mentioned, radioactivity cannot be slowed by any change in the temperature, pressure, or chemistry of the environment—only time can do this. The times of containment are long: Strontium-90 (^{90}Sr) and cesium-137 (^{137}Cs) require hundreds of years; plutonium (^{239}Pu) requires tens of thousands of years. Almost all the ^{90}Sr in a spent fuel assembly will have disintegrated after 500 years, but the same spent fuel assembly will contain more than a kilogram of ^{239}Pu after 24,000 years and will still be toxic after 400,000 years. Consider what has happened over the last 100,000 years: Civilization developed, there was at least one ice age, sea level rose and fell more than 100 meters, and there were thousands of earthquakes, volcanic eruptions, storms, and so forth. We have no sensory perception of nuclear radiation. We can sense heat associated with intense radioactivity, and we can see or hear instruments responding to it, but we can't see it, smell it, or feel it. To make matters worse, many of the offending elements in radwaste, such as iodine, strontium, and cesium, are biologically active as well as radioactive, and are readily incorporated and concentrated in organisms.

The objective in high-level radwaste disposal is to protect life on Earth for tens of thousands of years from inadvertent or uninformed contact with substances that could wipe out many terrestrial species, including our own. The critical requirements are as follows:

1. The site must be isolated from the biosphere.
2. The site must be in a geologically stable location.
3. The site must be safe from sabotage or illegal entry.
4. The site should not be in a potential resource area.

5. The site must be accessible by suitable, reliable means.
6. The site and disposal method must be economically feasible.

A repository for radwaste will be an arrangement of multiple natural and synthetic barriers to prevent invasion by water or organisms and escape by the radwaste. The first barrier will be to combine the wastes with other materials to form a relatively insoluble glass or ceramic solid solution, and then to enclose the mixture in sealed canisters of relatively inert metal and place the canisters in the repository. Over the last fifty years there have been eight serious suggestions for repository sites.

Salt Vaults. These have some important advantages. Salt beds have never encountered much water; they wouldn't exist if they had. Salt flows quickly to heal fractures and is a good thermal conductor. It is also relatively impermeable to diffusion. Salt beds are available in many parts of the world, including Kansas, New Mexico, Texas, Oklahoma, and New York State. Wastes buried in salt could be recovered with ease if we changed our minds. On the negative side, salt is a resource that might attract miners. Moreover, salt deposits do contain fluid inclusions that might burst and release corrosive brine, and salt is less dense than other sediments and therefore has an inherent tendency to form rising (buoyant) structures, called salt domes, which can push their way to the surface of the Earth.

Hard Rock Vaults. The principal objection to vaults in hard rocks, such as granite, is that such rocks tend to be fractured, even at considerable depth. It would thus be difficult to isolate a granite repository from groundwater. The most advanced preparations for a hard rock repository are under way in Sweden, where detailed surficial and subsurface surveys have defined a few bodies of crystalline rock that are sufficiently free of fractures. The Swedish plan is to encase the waste in copper canisters with 10-centimeter walls and pack these canisters in bentonite within a deep vault. Bentonite is a clay-rich material formed from volcanic ash that functions as a barrier because it absorbs moisture and detains many dissolved substances by adsorption.* Hard-rock vaults have been considered in the

---

*Absorption means the penetration (or assimilation) of some substance into the body of another. Adsorption means condensation of gas, liquid, or dissolved substances on the surface of a solid.

United States, but so far they have been less attractive than sites in bedded salt or volcanic tuff.

Volcanic Tuff. The United States is considering construction of a repository in bedded, volcanic tuff at Yucca Mountain, Nevada—an isolated site on the edge of the Nevada Test Site. Tuff has many advantages. It is highly absorbent of water; it flows to heal fractures, much like salt, but unlike salt is not soluble in water. When dry and compressed, tuff is more dense than salt and unlikely to flow upwards. Specific advantages of the Yucca Mountain site are that it is in an arid region; the water table is very deep (more than 500 meters); and the presence of interbedded, welded and nonwelded tuff deposits provides many structural and chemical advantages. Political advantages are also important: the test site has already been contaminated by radioactive materials and is far from any major concentration of voters. The most serious concerns about it are the possibilities of volcanic or tectonic disturbances in the future. The latter is especially worrisome and is being studied.

Deep Wells. We have learned from experience that injecting fluids under high pressure into deep rock formations can create instabilities that lead to earthquakes. Such was the result when the U.S. Army tried pumping chemical warfare wastes into deep wells in Colorado. Furthermore, the problems of ensuring isolation from groundwater are too great to consider.

Ice Caps. The advantages originally cited were that ice, like salt, is plastic, a good thermal conductor, and impervious to water. Unfortunately, ice caps are not stable; they flow. We have no idea what might happen if we created a slippery zone of wet ice on the bottom of an ice cap. It might slide into the sea. If that weren't enough to dissuade, the transportation criterion can not be satisfied. We might lose the waste en route to the repository.

Subduction Zones. This would seem to be an ideal solution, except that we don't really know much about the actual mechanics of subduction.

Abyssal Sea Floor. This is a serious possibility technically and almost impossible politically. Some of the most tectonically stable places on the Earth's surface are in the middle of the Pacific Ocean basin. Canistered wastes might be very safe buried in soft sediments in such places, but disposal at sea would be opposed by most nations on Earth. Again, transportation is a problem.

Space. Put it in a rocket and send it to the Sun! Great idea. The problems are that the transportation system is not sufficiently reliable and the cost is too high.

So what is going to happen? The United States has worried a lot about the problem of radwaste disposal, as have other nations. There is a general preference for deep burial on land, and different nations will choose different host rocks according to their national geology. The Swedes, Canadians, French, Swiss, and Japanese may choose hard rock sites. The United States may choose bedded tuffs, salt, or both.

Unlike material resources, which may be recycled, energy resources are absolute consumables. The energy we use to light our lamps, heat our homes, and move our automobiles can not be restored for any price. For two centuries, the developed world has benefitted from copious quantities of inexpensive, convenient fuel. In the century to come, as the lesser developed world achieves higher standards of living, as the fossil fuels become more scarce and more expensive, and as the by-products of combustion and nuclear fission become more and more burdensome, we will be obliged to seek new forms of energy and to use them with greater efficiency and with concern for their side effects. The energy sources are available. The challenge of the twenty-first century, like challenges of the past, will be to face the realities of the world and deal with them.

All prudent estimates of future consumption and remaining recoverable deposits of oil and natural gas indicate that those energy sources will be exhausted during the twenty-first century. Coal and indirect sources of oil and gas, such as oil-rich shale, will last longer, but increasing use of such fuels will cause dangerous environmental degradation. Inevitably, our grandchildren are going to face some tough challenges.

The use of radioactive materials and their disposal is a reality we must face now. Radioactive substances are a natural part of the Earth. Many of them are useful, with applications that go far beyond the energy and weapons areas and that are essential to our culture. Their use and disposal, however, require special precautions. Radioactive waste disposal, in particular, is an unavoidable challenge, so the question before us is not "should we?" but "what can we do about it?"

Should you and I be concerned about radwaste? Of course, we should, but in my opinion we and our descendants have more to fear from other toxic wastes, such as lawn and garden pesticides, which are more common in our everyday lives and often treated with careless disregard. To this statement I must attach one fearful exception: We have reason to be very worried about the illegal disposal of radioactive wastes by government agencies, particularly the military. During the 1950s the Soviet Union surreptitiously dumped 2.5 million curies of radioactive wastes into the Kara and Barents seas within the Arctic Circle. This criminal action is probably not unique.

To poison the waters of the Earth is to threaten the existence of life. Of all the resources of the planet, none is more essential. The elemental seeds of life may be scattered throughout the universe, but experience indicates that life can exist and evolve only on warm, wet rocks like Earth.

Moving masses of water, ice, and water vapor constantly warm Earth's high latitudes and elevations and cool the low latitudes and elevations. These movements constitute the *hydrologic cycle*. In the broadest terms, the hydrologic cycle is a balance between precipitation and evaporation on land and sea. From the oceans, 400,000 cubic kilometers of water evaporate each year, but only 370,000 return by direct precipitation. This mismatch is balanced by an excess of precipitation on land, where 60,000 cubic kilometers of water evaporate but 90,000 precipitate. The oceans do not dry up because the extra 30,000 cubic kilometers of water, which fall on the land as rain, snow, and sleet, are returned to the oceans by rivers and streams. Ninety-seven percent of the Earth's water is in the oceans. Two percent is in polar and glacial ice, including icebergs, and six-tenths of one percent is groundwater.* All the rest—

*These data do not include interstitial aqueous fluids in deeply buried rocks or water that is combined in crystal structures of hydrous minerals.

the lakes and streams, the rivers, swamps, clouds, and organisms—amount to less than one-tenth of one percent of Earth's 1.5 billion cubic kilometers of water.

For creatures who live on land, the critical water resource is fresh water, most of which is either frozen or underground. Surface waters, which are the most important water source for most land dwellers, comprise less than one-half of one percent of the Earth's fresh water supply. The small volume of surface waters is offset, fortunately, by relatively rapid rates of replenishment. Water stays in the atmosphere for about ten days. The residence time in rivers and streams is a few weeks, and in lakes and swamps it is a few years. The residence times of groundwater, however, range from decades to millennia.

Groundwater exists everywhere at some depth beneath Earth's surface. As you dig or drill into the ground, you pass through levels that may be quite damp but are not saturated with water. In this undersaturated zone there is air and open space. The water table is the top of the saturated region; below it all open spaces are filled by water. The amount of water in the saturated zone and the rate at which it moves depend upon the amount of pore space and how well the pores are connected to one another. The technical terms for these attributes are *porosity* and *permeability*. The water table is a mobile surface, which rises and falls according to the movement of water into and out of the ground. The surface is seldom flat and usually presents a subdued rendition of the overlying land surface. Where the water table intersects the Earth's surface one finds springs, lakes, rivers and streams, and, ultimately, the ocean.

An aquifer is a body of rock or sediment below the water table that is sufficiently porous and permeable to hold and transmit significant quantities of water. Gravel, sand, sandstone, and porous limestone make good aquifers, many of which are confined, above and below, by rocks that do not transmit water readily. Such rocks are referred to as *aquicludes*. In an artesian system groundwater within the aquifer is confined under pressure and therefore rises spontaneously in a well that penetrates the aquifer. Artesian systems are created when a sloping aquifer is confined by aquicludes.

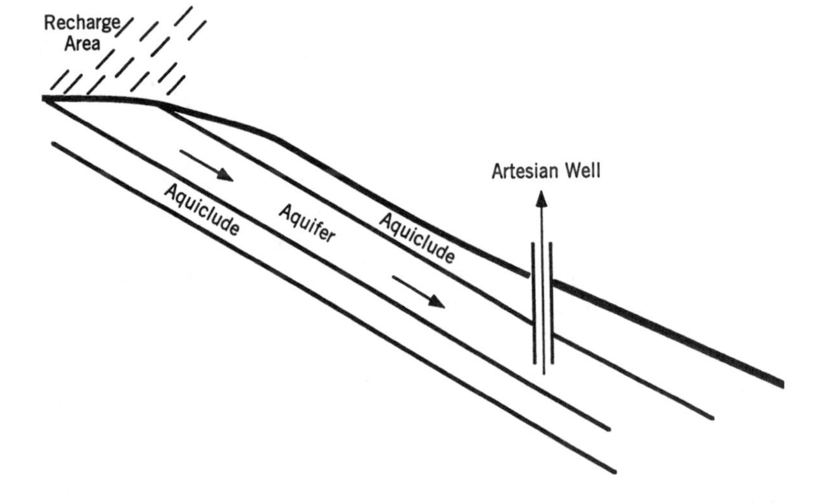

In 1856, Henry Darcy, a French engineer, prepared a report regarding the water supply of the City of Dijon. In the course of preparing his report Darcy performed experiments to better understand the flow of water through sand, which led him to a useful relationship known as *Darcy's Law*: The rate of flow through a porous, permeable medium is directly proportional to the permeability, cross-sectional area, and hydraulic gradient. The latter is the vertical distance traveled divided by the horizontal distance.* Permeability is a function of many factors, such as grain size and shape and degree of compaction. In general, permeabilities and flow rates are highest in unconsolidated gravel, clean sand, cavernous limestones, or fractured basalt. A typical flow rate is less than 2 feet per day.

We should regard water in aquifers as a nonrenewable resource because it flows so slowly through them. Pumping water from most aquifers is a form of mining, and over time the resource will be exhausted. A major example is the High Plains Aquifer of the United States, which is the principal source of water for 2 million

---

*Darcy's Law: $Q = K \times A \times (V/H)$, where Q is the flow rate, K is the coefficient of permeability, A is the cross-sectional area, V is the vertical distance, and H is the horizontal distance.

people spread across 174,000 square miles in Nebraska, Colorado, Kansas, Oklahoma, New Mexico, and Texas. There is one well for every square mile of this territory, each pumping water from the aquifer, and almost all of this water is used for irrigation. The region accounts for 20 percent of the acreage and 30 percent of the water used for irrigated crop growth in the United States.

Farmers began drawing water from the High Plains Aquifer in the late 1800s and over the next hundred years withdrew only 5 percent of the total available. Local depletions were significantly greater and increased from north to south. In Texas 23 percent of the water was drawn out during the first hundred years, whereas Nebraska experienced only an 8 percent decrease.[89] Although Nature can not recharge the aquifer as rapidly as people can (and will) draw water from it, the High Plains Aquifer will not soon run dry because it contains about 4,000 cubic kilometers (10^{15} gallons) of accessible water. Nevertheless, because pumping costs rise rapidly as the water level in the aquifer drops, dryland farming will replace crop irrigation over more and more of the High Plains. Texas may suffer effective loss of the aquifer before the year 2020; farmers further north will be able to pump water longer, although local conditions may cause individual wells to fail.

Another consequence of the low flow rates in most aquifers is their extreme vulnerability to long-term contamination. A polluted river, lake, or stream might be cleansed by enlightened public action within a few years, but this is not true of most aquifers. Many pollutants sink through the soil and unsaturated zones and slowly but surely enter the saturated groundwater zone. Once there, they will remain for hundreds or thousands of years.

Common groundwater pollutants are soluble nitrates from fertilizers and animal or industrial wastes. These nitrates are transformed into carcinogenic nitrites in our bodies, and their increased ingestion may result in an increase of gastrointestinal cancers. Nitrates from fertilizer and other sources are seeping into groundwaters all over the world. Under some circumstances, where free carbon is available and free oxygen is scarce, the dissolved nitrate may be reduced to harmless nitrogen and carbon dioxide gas, but this reaction is not sufficiently common to mitigate a growing problem. Furthermore, nitrates are just one of many unwanted byproducts of our crowded and careless world. From leaking landfills, buried fuel tanks, and myriad other sources, in spite of increased public concern and vigilance, the variety and concentration of pollutants in groundwater seems destined to grow.

In the twenty-first century water resources must be more tightly controlled and more intelligently managed than they are today. The responsibility of public and private organizations in industrialized countries to recover and treat their wastewater will become expected and accepted. Different grades of water will be assigned different applications. At the personal level very few of us will wash cars and water lawns with drinking water. At the global level world trade may be more influenced by water than by oil. Water-rich countries, such as Canada, Brazil, and Russia, may become major exporters of fresh water to water-poor countries, such as Egypt, India, and China, or to major per capita water consumers, such as the United States and western Europe. Consider that Canada has nearly twice as much fresh water as India, and on a per capita basis Canadians have two thousand times more fresh water than Indians.[90]

Water will also influence the world's economy by increasing its role as a source of energy. Rivers will continue to dominate hydroelectric power generation, but dramatic developments may occur in the oceans, where the power of tides and currents may, at last, be brought to bear upon our energy needs. Colossal construction costs, long construction times, inconvenient locations, changing flow conditions, and less expensive alternatives have traditionally worked against developers of tidal and current power, but if the technologies of hydrogen combustion or superconducting electrical power transmission come to pass, the enormous power resources of the ocean will become attractive indeed. Two developments, one in place, the other a dream, may foretell the future.

In 1966 the world's first major tide-powered hydroelectric station was commissioned on the Rance River in France. With its present net production of 500 gigawatt hours (GWh) per year,* La Rance remains the world's largest tide-powered facility, although it is tiny compared to the largest river-powered hydroelectric stations,

---

*Units are kilowatt hours (KWh) = 10^3 watt hours; megawatt hours (MWh) = 10^6 watt hours; gigawatt hours (GWh) = 10^9 watt hours; and terawatt hours (TWh) = 10^{12} watt hours.

which generate tens of terawatt hours (TWh) per year. Nevertheless, as a prototype La Rance may lead to more ambitious projects. Designs for a tide-powered generating station in the Severn Estuary on England's west coast, for example, predict annual outputs of 17 TWh, which equals 7 percent of the combined consumption of England and Wales.[91]

One of the most ambitious ideas of all was a suggestion, offered in 1981 by Spanish engineer Felix Cañada Guerrero, to place turbines in the Straits of Gibraltar, where *every second* strong currents transfer more than 2 million cubic meters of water back and forth between the Mediterranean Sea and the Atlantic Ocean.[92] Such turbines, he calculated, could generate more than 1,500 TWh per year, which would be more than twice the production of the world's largest river-powered hydroelectric plant, the Itaipú station on the Paraná River between Brazil and Paraguay. Perhaps Cañada Guerrero's visions could never overcome the many political, economic, and environmental obstacles they would face, but then again . . .

> Ah, but a man's reach should exceed his grasp,
> Or what's a heaven for?
>
> *Robert Browning*

Ten

Diamonds

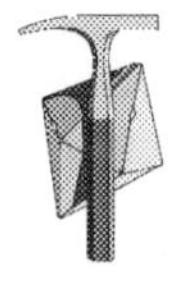

Supremacy in the mineral kingdom belongs to diamond. No mineral is harder, none is more pure, none conducts heat so well, and none is more brilliant. Diamond is an important commercial commodity, contributing to the economies of nations on four continents as an exploited mineral and supporting manufacturers around the world as a synthetic product of proven performance and increasing potential.

In 1993 more than 57 million carats* of gem-grade diamonds were produced from mines in 22 nations. Of these Australia was the most productive, followed by Botswana, Zaire, Russia, and South Africa. These same mines also produced more than 50 million carats of industrial diamonds. At the same time more than 450 million carats of synthetic diamonds were produced by plants in 14 nations. The leading producers of synthetic industrial diamonds are in the United States, Russia, South Africa, Ireland, and Japan.

The overall volume of diamond commerce exceeds $7.5 billion annually. Gems, with prices determined by monopolistic control of markets, remain by far the big money makers, but industrial applications, particularly of synthetic diamond, are increasing rapidly. The competitiveness—if not superiority—of synthetic over natural industrial diamonds exists because precise control of the physical conditions of crystal growth and of the trace element composition of the

---

*Seeds of the carob tree (*ceratonia silique*) were once used to measure the weights of pearls. The formal unit, one carat, comes from those seeds and equals 200 milligrams or about 0.007 ounces.

synthetic diamond allows the manufacturer to offer a product that performs consistently and is precisely suited for specific applications.

Leading 1993 Diamond Producers[93]
(millions of carats)

	Natural		*Synthetic*
	Gems	***Industrial***	***Industrial***
Australia	19.0	23.2	0.0
Botswana	12.0	5.0	0.0
China	0.23	0.85	15.5
Ireland	0.00	0.00	66.0
Japan	0.00	0.00	32.0
Russia	8.0	8.0	80.0
South Africa	4.3	6.05	75.0
Sweden	0.0	0.0	25.0
Ukraine	0.0	0.0	10.0
United States	0.0	0.0	103.0
Zaire	9.5	5.5	0.0
World total	57.205	50.390	456.000

Determining the composition of diamond was an early achievement of quantitative chemistry. On the basis of its optical properties Isaac Newton suggested that diamond was "probably an unctuous [oily] substance coagulated." In 1694, at the Accademio del Cimento in Florence, scientists "vaporized" a diamond by focusing sunlight on it. In fact, they burned the diamond, but the flame was virtually invisible in the bright sunlight so the diamond seemed simply to vanish into thin air, which must have been a surprise. Then in 1772 Lavoisier focused sunlight on diamonds in a bell jar and produced gas (carbon dioxide) similar to that liberated by fermentation or by burning charcoal. Twenty-five years later Smithson Tennant, who with James Hutton had studied chemistry under Joseph Black in Edinburgh, burned a weighed quantity of diamond and generated the exact same amount of carbon dioxide ("fixed air") that Lavoisier had produced from an equivalent mass of charcoal.

Diamonds are crystals of carbon, which is the fourth most abundant element in the solar system. More than 90 percent of all known chemical compounds contain carbon, which is the lightest element in Group Four of the Periodic Table. Like silicon, another Group-Four element, carbon is able to form four covalent bonds, but its electronic versatility is greater than that of silicon, allowing a wider variety of bondings with itself and other elements.

Diamond and graphite, the two most common forms of elemental carbon, are the original *allotropes*, a word coined by the great Swedish chemist Jöns Jakob Berzelius (1779–1848) to indicate different forms of the same element. The differences between graphite and diamond are due to the way the carbon atoms are joined together. In graphite the atoms are arranged in sheets of hexagonal rings, within each of which the carbon atom is firmly bonded to three others. The sheets are stacked one on top of another, but not firmly bonded together; they are easily separated, making graphite flaky, soft, and slippery. In crystals of diamond each carbon atom is firmly bonded to four others disposed toward the corners of a tetrahedron. This three-dimensional structure is very strong and gives diamond special qualities, such as its extraordinary hardness.

Diamond is the hardest substance known. Diamonds can scratch anything, which is why they make good cutting tools for all kinds of things, such as crystals, rocks, ceramics, and hard metals. Hardness alone, however, does not explain why diamond is a precious gem. The beauty of diamond crystals derives from the way they transmit and reflect light.

The velocity at which a substance transmits light is expressed as its *refractive index*—the ratio of the velocity of light passing through a vacuum to the velocity of light passing through the substance:

$$\text{Refractive index} = \frac{\text{velocity of light through vacuum}}{\text{velocity of light through substance}}$$

Diamond has a very high refractive index. Light travels 186,000 miles per second through a vacuum but only 77,500 miles per second through a diamond. Light travels almost twice as fast through quartz as it does through diamond.

The ratio of the intensity of light reflected by a crystal to the intensity of the light shining (incident) upon the crystal is called the *reflectivity*:

$$\text{Reflectivity} = \frac{\text{intensity of reflected light}}{\text{intensity of incident light}}$$

Reflectivity is related to the refractive indices of the reflecting substance and the refractive index of the surrounding medium. The general relationship for a substance like diamond, which transmits light at the same velocity in all directions, is

$$\text{Reflectivity} = \frac{(n - n_o)^2}{(n + n_o)^2}$$

Where n is the refractive index of the substance and n_o is the refractive index of the surrounding medium. The refractive index of air is 1.00029. Some comparative data for reflectivities in air are shown in the following table. For a transparent substance, diamond is extraordinarily reflective. Substances with refractive indices and reflectivities higher than diamond are usually deeply colored, almost opaque.

Refractive Index	*Reflectivity in Air*	*Substance*
1.2	0.8%	
		Water and Ice
1.4	3%	
		Fluorite
		Halite
		Quartz
		Emerald
1.6	5%	
		Garnet
		Ruby
1.8	8%	
2.0	10%	
		Zircon
2.2	14%	
2.4	17%	
		Diamond
2.6	20%	

Refractive index and reflectivity are not the only reasons for diamond's beauty, however. Another important factor is how the velocity of light in diamond changes with the color, or wavelength. This property, which gives diamond its fire, is referred to as the dispersion of the refractive index. A substance with high dispersion acts like a prism and breaks white light into its constituent colors because the different colors are transmitted at different velocities and are therefore bent by different angles when they enter and leave the substance. All translucent substances do this, but the effects are most striking with crystals of high dispersion. The dispersion of the refractive index is defined as the difference between refractive indices for light of specific wavelengths. For convenience we often use two bright lines in the hydrogen spectrum: a blue line (wave length = 4861 Å) and a red line (wave length = 6563 Å)*. As the following data indicate, diamond shows very high dispersion, especially for such a transparent substance. Zircon and hematite, which show comparable dispersions, are often nearly opaque.

Substance	***Dispersion***
Fluorite	0.0045
Ice	0.0062
Quartz	0.0078
Ruby	0.011
Emerald	0.013
Peridot	0.013
Zircon	0.023
Diamond	0.025
Hematite	0.5

Diamonds make superb gems because they are limpid yet exceedingly hard, refractive, reflective, and strongly dispersive of white light.

*Named for Swedish physicist Anders Jonas (1814–1874), the angstrom (Å) equals 10^{-10} meters.

If to far India's coast we sail,
Thy eyes are seen in di'monds bright

John Gay

Where do diamonds come from? Had you asked the elegant seventeenth-century French jeweler, Jean Baptist Tavernier, he would have said the Orient. Tavernier made half a dozen trips to Asia, beginning in 1631, to buy and sell gems. He handled many legendary stones, including the Blue Tavernier diamond, of which the Hope diamond is thought to be the largest part.

Prior to 1870 diamonds were known to occur only in stream gravels, where they are concentrated along with other relatively dense minerals. Alluvial diamond deposits were known in India as early as 800 B.C. At the beginning of the eighteenth-century gold miners in Minas Gerais, Brazil, found diamonds but did not recognize them as such. They apparently used the distinctive pebbles as poker chips. Eventually the pebbles were identified as diamonds, but even then people doubted that they could have originated in Brazil, so strong was the notion that diamonds were oriental.

The association of diamonds with gold in placer deposits was also confusing. Because of this association diamonds were thought to be genetically related to gold, but this is not exactly true. Diamond and gold may derive from the same geologic terrain, but they occur together in stream sediments only because flowing waters winnow their cargo, concentrating together grains of like density. Unlike gold, which forms in a variety of geologic situations, most natural diamonds originate in similar circumstances, and these circumstances involve not the meandering streams of India or Brazil but deep-seated igneous injections into ancient continental crust. These injections were first identified in southern Africa.

In December 1487 a gale swept two Portuguese caravels commanded by Bartholomeu Dias around the southern tip of Africa, giving Dias priority among European explorers—though perhaps not among ancient Phoenician mariners—for rounding the

Cape of Good Hope. After sailing 500 miles farther eastward along the coast the crew forced Dias to turn back, and the two ships returned home, passing within sight of the Cape, which they named *Cabo de Tortmentoso*, the Cape of Storms. Dom João II, King of Portugal, hoping that this newfound passage might circumvent Muslim and other barriers to Asia, renamed it *Cabo de Boa Esperanza*, the Cape of Good Hope. Ten years later, in November of 1497, Vasco da Gama rounded the Cape on his way to India. From then until today Table Bay, tucked in behind Table Mountain, the Cape's 35,000-foot sandstone rampart, has been a refuge for storm-tossed sailors passing between the Atlantic and Indian Oceans.

In 1652 the Dutch East India Company sent Jan van Riebeeck with 90 men to establish a revictualing station at Table Bay. Eventually the demand for grain, meat, and vegetables induced the company to allow settlers to establish farms beyond the original station, incursions that were resisted by the indigenous peoples, the Khoikhoi and the San—known to Europeans, respectively, as Hottentots and Bushmen. By the mid-eighteenth century the Khoikhoi resistance was crushed, and by the mid-nineteenth century the San had been decimated and driven into the Kalahari desert. Settlers, mainly Dutch Boers, extended the borders of the Cape colony north and east until they encountered formidable opposition by the Zulus and other Bantu-speaking peoples.

Dutch control of southern Africa ended at the end of the eighteenth century. The Dutch East India Company declared bankruptcy, and Britain, at war with France, seized the Cape colony in 1795. Britain's interest in the Cape, until the discovery of diamonds and gold, was as a link in the Royal Navy's global hegemony. As British control tightened over the Cape, Boer families trekked north across the desolate Great Karoo and up onto the interior plains, the high veld, drained by the Orange and Vaal Rivers, to find new homes.

Among the people of the high veld were many of mixed race. About them an early traveler told an interesting tale:

> The people in this part, being a mixed race, went by the name of Bastaards; but having represented to the principal persons the offensiveness of the word to an English or Dutch ear, they resolved to assume some other name. On consulting among them-selves, they found the majority

> were descended from a person of the name of Griqua, and they resolved hereafter to be called Griquas.[94]

And so the land became known as Griqualand, and it was in this region that diamonds were discovered.

The first diamond known to Europeans in South Africa was a 5-carat stone purchased in 1859 by a Boer clergyman from a native who probably found it in gravels of the Vaal River. The first diamond actually found by Europeans in South Africa was the Eureka diamond. This bluish-white stone, which weighed just over 21 carats, turned up in the dirt of a farm near Hope Town, a village on the south bank of the Orange River in 1866 or '67. In this village a man named Jacobs worked and lived on a property known as the De Kalk farm. One day Mrs. Jacobs sent her fifteen-year-old son Erasmus and his friend, a Khoikhoi boy named Klonkie, out to cut a branch that she might use to clean a clogged drain. Erasmus came home with a curious pebble, which Klonkie had found and given him. About the size of a gambler's die, it was surprisingly dense and sparkled in the sunlight. Mrs. Jacobs showed the stone to Schalk van Niekerk, another resident of the De Kalk farm, who was a "rockhound." Van Niekerk had seen pictures of raw diamonds in books, and this stone looked like the pictures. He went to the window and drew the stone along the glass, leaving a deep scratch—which must have pleased Mrs. Jacobs no end. Eventually the stone was confirmed to be a diamond and was purchased by the Cape governor, Sir Philip Wodehouse. Van Niekerk received £350 (almost $117,000 in 1995)*, which he split with Mrs. Jacobs. When cut, the Eureka weighed 10.73 carats.

Two years later a Griqua boy named Booi found a much larger diamond while herding his sheep across another farm near Hope Town. Booi needed a place to spend the night, so he asked farmer Duvenhage if he might stay at his place in exchange for the pretty stone. Duvenhage allowed as how he was "no van Niekerk" and threw the boy out. When van Niekerk met Booi a day or so later he gave the boy his horse, 10 oxen, a wagon, and 500 sheep in exchange for the stone. Van Niekerk sold the stone, which weighed 83.5 carats, to a group of traders for £11,300 (almost $3.8 million in 1995). It was later sold in London for £25,000 (more than $8.3 mil-

---

*1£ ≈ $4.86 in 1867. $1 in 1867 ≈ $68.47 in 1995.

lion in 1995) and became known as the Star of South Africa. When it was brought before the House of Parliament the Colonial Secretary proclaimed, quite accurately: "This diamond, gentlemen, is the rock upon which the future success of South Africa will be built."[95]

The geological origin of diamonds was eventually revealed when a wagoner named Bam found a diamond near the Riet River in 1870. Bam and some friends returned to the spot where he had found the stone and began to dig. At first they did not know that they were digging into the Koffiefontein diamond pipe, but before long Bam and the world knew that alluvial diamonds are derived by weathering and erosion from the cylindrical igneous intrusions that are known today as *kimberlite pipes*.

The rock called kimberlite is similar to peridotite, the olivine-rich rock that is thought to dominate the upper mantle of Earth. But unlike the peridotites that are the unmelted residues of a partial fusion process, or those that are accumulations of olivine crystals due to fractional crystallization, kimberlites are not residues or fractions. They crystallize from magma of their own composition, a substance rich in the elemental constituents of olivine ($(Mg,Fe)_2SiO_4$) but also distinguished by noticeably high concentrations of potassium, chromium, calcium, and carbon. Deep in the upper mantle the carbon is thought to exist as calcium and/or magnesium carbonate and as diamond, but when the magma approaches the surface much of the carbon is converted to carbon dioxide and graphite.

Kimberlite magmas originate more than 300 kilometers below the surface. From such depths they probably flow slowly upwards, as low density blobs, displacing denser mantle beneath the thick continental lithosphere. At the base of the lithosphere, which may be 200 kilometers deep, they stop rising. For them to rise further the lithosphere must crack so that the kimberlite fluid can worm its way up fissures toward the surface. The mechanism for such fracturing is not understood. It may be caused by local or regional stresses, but no one knows. In any case it happens, and the kimberlite fluid rises. Within at least 5 kilometers of the surface the fluid begins to boil off carbon dioxide. This transforms it into a churning, gas-charged, semisolid substance that drills its way upward until it explodes into the atmosphere. Rocks are ripped from the walls of the conduit, ground to bits, and carried along in the rising stream. These accidental additives, called *xenoliths* and *xenocrysts*, add substantial quantities of crustal and mantle material to the kimberlite. The walls of the conduit, as well as the xenoliths and xenocrysts, are abraided and smoothed to streamlined, aerodynamic forms.

Once the crust has been pierced, the gas expands explosively and the entire column, extending down hundreds of kilometers into the Earth, accelerates upward. No one has ever seen a kimberlite eruption, and the erupted material is so fragmented that surface accumulations (i.e., volcanoes) are not long preserved. For me these events evoke the image of some subterranean fiend firing a howitzer straight up through crust, and the image may be apt. In the nineteen seventies Tom McGetchin, a bright young volcanologist from MIT with some ballistics experience at the Air Force Academy, measured the sizes of rocks thrown out from kimberlite vents and the distances the rocks had been thrown. He analyzed these data as a ballistics problem and arrived at an amazing model of kimberlite eruption:[96]

Depth below the Surface	***Velocity***	***Gas Temperature***
50 kilometers	25 meters/second	1,000°C
1 kilometer	116 meters/second	661°C
300 meters	186 meters/second	440°C
Ground level	334 meters/second	–19°C

McGetchin's model showed the erupting kimberlite magma accelerating from 56 miles per hour near the base of the crust to 747 miles per hour, which is nearly the speed of sound, when it reached Earth's surface. Just as expanding gas cools when it leaves an aerosol spray can, the gas in the erupting magma cooled rapidly as it expanded, cooling the entrained solids until the rising magma reached within about 3 kilometers of the surface. At that depth the temperature was about 870 degrees, and from there on up the gas expanded and cooled too rapidly to equilibrate with the rocks, which were still around 800 degrees when they reached the surface. The rapid ascent and cooling of the material in McGetchen's model helps partly to explain how rocks and crystals, including diamonds, can survive the extreme decompression from the deep upper mantle, where pressures exceed 50 kilobars, to the surface at a pressure of 1 atmosphere.

Diamond pipes don't occur just anywhere on Earth's surface. They are never found in the ocean basins, perhaps because the Earth's lithosphere is too thin—the thermal gradient (temperature versus pressure or depth) is too hot—beneath the oceans. Diamonds rising through the suboceanic mantle would be converted to graphite, which is the stable form of carbon at high temperature and low pressure. Instead, they form in mantle deep beneath ancient continental crust, such as one finds in the shield areas of Africa, South America, or Siberia, and they do not enter shallower depths until eruption. Continental shields are old surfaces that have been eroded down to rather flat plains. In such terrains the diamond pipes don't normally show up as dramatic topographic features, prominent hills, or holes. Not, that is, until the diggers get to work! Because the diamonds are confined to the kimberlite intrusion, which is generally softer than the rocks of the surrounding country, the miners dig straight down into the pipe. The holes become very deep very quickly. Before long they become too deep and too steep-sided to work, and underground mining operations take over.

The normal concentration of diamond crystals in kimberlite is less than 100 parts per billion, so it is necessary to mine vast amounts of rock. Separating the diamonds from the ore must also be performed on a large scale, and it must be highly effective. When the kimberlite comes out of the mine it is crushed and sieved, during which about a quarter of the material is lost as fine sand and mud.

In the traditional separation process the crushed material then went through a series of density separations by floating or sinking in various dense liquids. The density of diamond is 3.5 grams per cubic centimeter, which is more than most of the minerals in kimberlite. The diamonds and other dense minerals, such as garnet, sank in the dense liquids while the lighter materials, such as olivine (3.3 g/cm^3), floated. After this separation the diamond-bearing concentrate was reduced to less than 2 percent of the original material. Next the concentrate was put through magnetic separators. (Diamonds are not attracted by magnets, but other dense minerals, such as garnets and iron oxides, are.) The diamond concentrate was then no more than 1 percent of the original ore. The next step in the traditional process was to pour the diamond concentrate, in a stream of water, over a flight of grease-covered steps. Most diamonds are not wetted by water, so they stuck to the grease while the other materials were washed away. Large diamonds were picked from the grease by hand. Periodically the grease was scraped off and melted. The diamond concentrate was then filtered off.

In the 1980s the traditional process of diamond separation was replaced by a technology invented in the Soviet Union that relies upon a phenomenon called X ray luminescence. Diamonds glow when exposed to X-rays; most other minerals do not. In the X-ray sorter a stream of diamond-bearing concentrate falls through total darkness across an X-ray beam. The point of intersection is monitored by a light detector, and if a diamond flickers the light detector triggers a blast of compressed air that deflects the falling diamond into a special chute. After washing and drying the diamonds go to the sorting house, where they are sorted according to size, color, and perfection.

The largest diamond ever found was the Cullinan Diamond, which weighed almost a pound and a half (3,106 carats). It was cut into 105 stones, the largest of which, weighing 530 carats, is in the British Royal Sceptre. The Cullinan Diamond was found on January 25, 1905, at the Premier Mine near modern Kimberly, by the surface manager, Daddy Wells. Actually, it was found by a miner who saw it glittering on the quarry face in the late afternoon sun and brought the stone, which was about the size of a large potato, to the manager's shack. Someone in the shack snorted that, of course, it wasn't a diamond and threw it out the window. The miner began to bellow, attracting Daddy Wells' attention. He and the miner went back out and retrieved the stone, which Wells recognized immediately.

Diamonds are usually found as individuals—that is, as unattached crystals embedded in kimberlite—but studies of rock fragments that have been carried up by the kimberlite indicate that many diamonds originate in eclogites—mantle rocks composed mostly of garnet and pyroxene minerals. They also occur in peridotites, and some may form directly from the kimberlite magma while it lies deep beneath the lithosphere.

Diamond pipes are known to have been emplaced in Earth's crust over a period of at least 1.5 billion years, which is approximately one-third of Earth's history. The oldest known emplacement is the Kuruman pipe in South Africa, which is 1.6 billion years old. The most recent, also in Africa, are pipes in Namibia and Tanzania that erupted only 50 million years ago. Young diamond pipes have also been found in Canada's Northwest Territories.

The age of the diamonds in a pipe is not necessarily the same as the age of the igneous intrusion. One of the first to discover this fact was a graduate student at MIT named Steve Richardson, and his road to discovery was hair-raising.

Dust-sized crystals of other minerals are often found within diamonds. Although these microscopic inclusions spoil the gem value of a stone, their scientific value is great because they represent some of the other minerals that were present deep in the Earth where the diamond formed, and they provide a means for determining the absolute ages of the diamond host crystals. The age dating technique is tedious. Under a binocular microscope one cracks the diamond open like a nut, picks the dust-like inclusion up on the tip of a fine needle, and places it in a vial. After doing this for many months one may have collected enough grains to measure the relative concentrations of radioactive and radiogenic isotopes (e.g., ^{235}U and ^{207}Pb) in the collection of inclusions.

Richardson spent many months gathering garnet dust inclusions from several hundred diamonds that had been culled for him from perhaps as many as a million stones by diamond sorters in South Africa. When he had enough material, he processed the dust through the intricate and delicate chemical treatments required to separate and concentrate the elements of interest. Then one night, using a tiny, hand-held teflon pipette, he transferred his precious sample solution from its beaker to the mass spectrometer filament, evacuated the chamber, switched power to the filament and high voltage to the source lens, and waited for the first ions to hit the detector. At that moment a general power failure swept across the City of Cambridge, and the laboratory and its mass spectrometer were suddenly dark and silent. Sitting beside the still machine, Richardson closed the high vacuum valve and telephoned his advisor, Stan Hart:

"Steve here," he said quietly. "Stan, there's been a power failure. My stuff's in the mass spec, but the system crashed. What should I do?"

How can he be so calm? Hart wondered.

"Well, Steve . . . There's not much you can do. . . . Turn the current down so a surge doesn't blow your sample off the filament when the power comes back. . . . Wait and hope the machine comes up okay when the power's back on."

"Oh."

"Good luck. . . . Call and let me know what happens."

"Right. . . . Bye."

Alone in total darkness, Steve waited beside the machine. Seconds, minutes, hours? No, it only seemed forever. Then, as suddenly as before, the lab was bright. Lights twinkled on the mass spectrometer, and as calm as its young operator, it returned to life; the filament glowed, ions flew down the tube, detectors responded, and data—good data—poured out.

The radiometric ages thus obtained indicated that many diamonds were much older than their host kimberlites. Some had existed in the Earth for billions of years before being carried to the surface by younger kimberlite magmas. In addition to confirming that the diamonds had formed independently of the kimberlite magma, the ages also gave strong support to the idea that the mantle beneath the ancient crust of southern Africa is itself ancient and has been unusually cold, cold enough to preserve diamond for billions of years.

Where does the carbon come from in the first place? The nature of the carbon atoms holds some clues to the answer to this question. All carbon atoms have six protons, but they may have six, seven or eight neutrons, meaning that a carbon atom may have an atomic weight of 12, 13, or 14. Carbon-14 (^{14}C), a short-lived radioactive isotope that forms in Earth's atmosphere, is useful for dating archæological artifacts and other young materials, but it has not been detected in natural diamonds. Carbon-12 (^{12}C) and carbon-13 (^{13}C) are stable isotopes present in varying relative amounts in different situations, so the ratio $^{12}C/^{13}C$ may indicate the origin of carbon in a substance. The ratios for diamonds indicate that they come from two sources. One group apparently comes from Earth's interior and is made of carbon that has never been at Earth's surface; the other group contains carbon that had been at the surface of the Earth and was returned to Earth's mantle by subduction.

As soon as people knew what diamonds were made of, in the early nineteenth century, they tried to make them. Prior to the discovery of the kimberlite pipes in South Africa all attempts at diamond synthesis were performed at ordinary pressures and involved either precipitation from some carbon-rich solution, electrolysis of some carbon compound, or injection of carbon or carbon compounds into an electric arc. However, no confirmed diamond synthesis resulted from these early experiments.

The discovery of the South African diamond pipes led to the realization that diamonds are stable only at high pressures. This insight was an essential step in discovering how to make them. One of the earliest high-pressure experimenters, James Hannay of Glasgow, filled thick-walled wrought iron tubes with hydrocarbons, welded the tubes shut, and heated them to glowing red in a furnace. This exceedingly dangerous procedure resulted in a succession of furnace-smashing explosions—all but 3 of his 80 experiments blew up—but miraculously no one was killed. In 1880 Hannay produced pieces of a transparent substance that Professor Nevil Story-Maskelyne of Oxford, the grandson of the man who first measured Earth's mass, identified as diamond. And diamond it was, although subsequent analyses have shown it to be fragments of natural diamonds. Hannay, or someone, had salted the experiment.

In the 1890s a more reputable scientist, Professor Henri Moissan of Paris, who won the 1906 Nobel Prize in Chemistry, claimed to synthesize diamonds by dissolving carbon in molten iron and suddenly quenching the molten metal. The quenched iron was then dissolved in acid and the insoluble residue examined for diamond. Crystals were found that had the physical and chemical attributes of diamond. Unfortunately, however, these crystals have been lost and cannot now be subjected to modern analysis, and careful attempts to reproduce the experiments have never yielded diamonds. After Moissan's death Mme. Moissan is reported to have suggested that a laboratory assistant had added diamond chips to the experimental residues "in order to please the old man and to avoid the tedium . . ."[97]

The nineteenth-century experimentalists were attempting diamond synthesis under pressure conditions that were, unbeknownst to them, far from the stability field of diamond. We know now that the pressure-temperature line between the stability fields of graphite and diamond slopes up from around 15 kilobars at room temperature to around 100 kilobars at 3,000 degrees. In order to synthesize diamond one would have to exceed the pressure conditions along that line.

More than anyone else, the person who developed the apparatus and techniques needed to reach the diamond stability field was Percy Bridgeman, a lifelong Bostonian, Harvard graduate, and Harvard professor who was awarded the 1946 Nobel Prize in Physics for his achievements in high-pressure science and technology. In 1905, when Bridgeman was a twenty-three-year-old graduate student, the most advanced laboratory techniques permitted experiments at 3 kilobars but no more. While using these techniques to study optical phenomena at high pressures, Bridgeman inadvertently broke his apparatus. As he rebuilt this equipment he invented devices and techniques that enabled him to hold an experimental charge at 20 kilobars and started him on a progression of technical improvements that led to pressures approaching 500 kilobars. During his career Bridgeman made several attempts to synthesize diamond. At

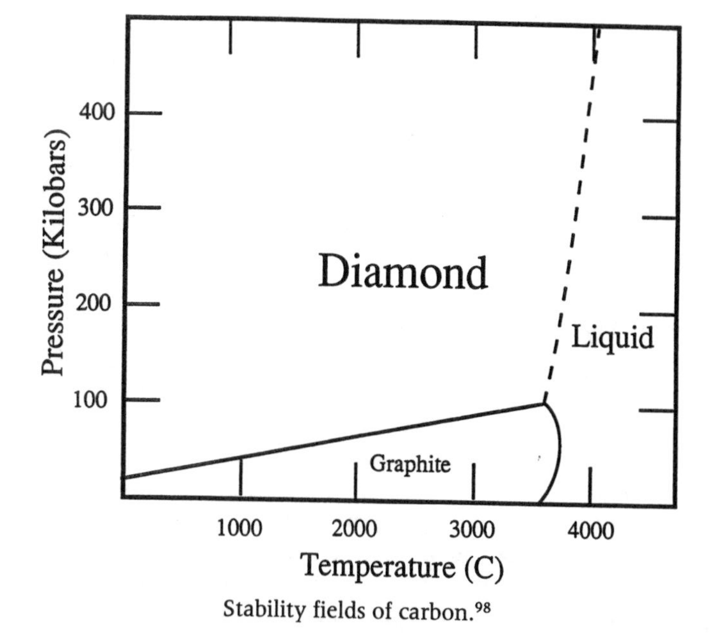

Stability fields of carbon.[98]

room temperature he subjected graphite charges to 425 kilobars; at red heat he applied pressures of 70 kilobars; and using thermite* as an internal heat source he attained 3,000 degrees centigrade at 30 kilobars. But it was all in vain. Although he was able to enter the diamond stability field, without a catalyst the graphite to diamond reaction would not proceed fast enough to produce visible results.

The bonds between carbon atoms are exceptionally strong. Even though a crystal of diamond or graphite may be in a physical situation that is well outside its pressure-temperature stability field, it will persist because of the enormous energy required to break the carbon–carbon bonds before the crystal structure can change. This explains the stubborn persistence of the diamond on your finger or of graphite compressed by more than 30 kilobars.

The first confirmed synthesis of diamond was achieved by scientists in Sweden on February 15, 1953. The Swedish electrical company ASEA had been attempting diamond synthesis off and on for about a decade and had developed a spherical high-pressure device that could reach 75 kilobars. Within the sphere a thermite reaction could raise the temperature of the compressed experimental charge to more than 2,000 degrees. During the experiment the charge would remain at 75 kilobars and move deeper and deeper into the diamond stability field as it cooled. As Bridgeman and others had found, however, experiments with pure carbon did not yield diamonds. They needed a catalyst to speed the rate of reaction, and the first successful one was iron. Rather than starting with pure carbon the Swedes started with iron carbide and graphite, which during the experiment were transformed into a solution of molten iron with dissolved carbon. As the charge cooled about 40 small diamond crystals precipitated out of the molten iron.

Within a year, on December 16, 1954, a team of scientists at the General Electric laboratories in Schenectedy New York, led by H. Tracy Hall, also synthesized diamonds by adding iron to the experimental charge. General Electric announced its achievement in Feb-

---

*Invented by Claude Vauten of London in 1900, thermite is a mixture of aluminum metal and a metal oxide, such as iron oxide or chromium oxide, that produces intense heat upon reaction. Thermite is used to weld large metal objects, such as pipes or rails. It is also used in munitions.

ruary of 1955 and moved quickly to establish synthetic diamond as a commercial product that could be modified in various ways to perform as well or better than natural diamond as an abrasive. Larger, gem-quality diamonds were also synthesized, but producing them requires such extraordinary efforts that they have not been competitive with the natural stones.

Much is known about growing diamonds in various colors, shapes, and sizes, and much has been learned about the range of pressures and temperatures over which diamond is stable. Although the experimental data are still few and imperfect, they indicate that the diamond structure is extraordinarily robust, persisting to pressures in excess of 10 megabars and temperatures well above 5,000°C. The data indicate that the stable form of elemental carbon is diamond throughout the deepest regions of Earth and throughout the carbon-rich giant planets that lie far beyond Earth's orbit.

Eleven

Neighbors Near the Sun

Aristarchos of Samos (310 B.C.–230 B.C.) first gauged the cosmos with geometrical logic and quantitative observations. One of his greatest achievements was his audacious attempt to measure the relative distances between the Earth, the Moon, and the Sun. Aristarchos realized that when Earth observes a half moon, the plane of the terminator on the Moon is normal to the Moon–Sun line. By measuring the angle between the Earth–Sun and Earth–Moon lines at such times, he could draw a right triangle, with the Earth, Sun, and Moon at each corner. His results were greatly in error, but he demonstrated that relative distances could, in principle, be measured in space. Other observers would follow his lead.

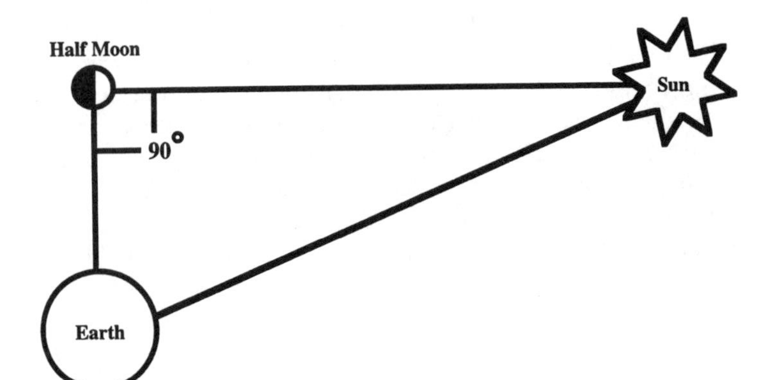

In the summer of 1609, Galileo discovered that the Moon has two crustal terrains, light and dark, which he called *terrae* and *maria,* or lands and seas.

Isaac Newton, through consideration of Earth's oceanic tides, concluded that the mass of the Moon was about 2.5 percent that of Earth. Laplace, using several different approaches, including Sir Isaac's tidal model, estimated the mass at slightly less than 1.5 percent that of Earth. The modern value is about 1.2 percent of Earth's mass.

By sighting the Moon simultaneously from widely separated observatories on Earth, it is possible to determine how far away and how big it is. Such observations and calculations date back to ancient times, but were hampered by imprecise instruments for measuring time as well as space. In the mid-eighteenth century, European astronomers used sightings from Greenwich, Paris, Berlin, and the Cape of Good Hope to determine these dimensions. The modern value of the mean distance to the Moon is 384,400 kilometers, and the lunar radius is 1,738 kilometers.

Earth is 50 times bigger than the Moon, but 83 times more massive because the density of the Moon (3.34 g/cm^3) is significantly less than the density of Earth (5.5 g/cm^3). The similarity of the Moon's density and the density of Earth's outer mantle is striking, and probably no coincidence.

The Moon is moving slowly away from the Earth as tidal movements of the ocean slow the Earth's rotation and cause angular momentum to be transferred to the Moon. The rate of escape is not constant, but is strongly affected by the area covered by shallow seas, which has varied a lot during Earth's history. At some time in the past the Moon may have been as close as the Roche limit,* where it would break apart, but we cannot yet define that time.

One important fact, which we did not fully appreciate before going to the Moon, was the frequency of solid-body collisions in the solar system, especially during its early history. Because the evidence of meteorite impacts on Earth is quickly destroyed by erosion, we underestimated the importance of this phenomenon. But we know now, from observations and from computer models, that col-

---

*Roche limit: "The critical distance inside which an idealized satellite with no tensile strength would shatter due to tidal forces exerted by its parent planet."[99]

lisions have been frequent, and some have been events of great consequence for Earth's physical and biological evolution.

The history of lunar exploration and lunar understanding took off, literally, in January 1959, just fourteen months after *Sputnik,* when the Soviet Union made the first lunar fly-by with *Luna I.* Eight months later, the Soviets put *Luna II* on the Moon, and one month after that *Luna III* sent back the first pictures of the lunar far side. Something that is often overlooked is that the Soviets led the way to the Moon. For all its faults and failings, the Soviet Union was a world leader in many areas of science and technology. The pre-Apollo imagers and robots taught us many things about the lunar surface. The most important, for understanding the origin of the Moon, was that its maria are covered by basalt, the most common lava on Earth.

At 10:56:20 P.M. on Sunday, July 20, 1969, Neil Armstrong stepped down from the *Apollo XI* lunar module and scooped up some lunar soil. Since that moment the study of these precious moon rocks and soils has been a source of some amusement and misunderstanding. We did not go to the Moon to collect rocks. We went to catch up with, and surpass, the Russians in space technology and to enhance our nation's prestige in the world. The lunar samples were an exciting by-product of the effort.

The astronauts brought back 21.5 kilograms of lunar material, of which 200 grams were split into small gifts that President Nixon distributed; 900 grams went on display in the United States and overseas; 700 grams were consumed in precautionary biological tests; 7 kilograms were distributed among 150 principal investigators and their associates in laboratories throughout the United States and the noncommunist world; and the remainder was stored away for future study. The samples were distributed to the investigators in early September. No publication of results was permitted until the entire group of scientists met in Houston in January 1970.

At the Smithsonian Astrophysical Observatory in Cambridge, Massachusetts, John Wood had decided, beforehand, to study the "coarse fine-grained" fraction of the lunar soil. He wanted particles that would be large enough to contain several phases—in other words, he wanted rocks, not minerals—but he wanted small ones. He reasoned, correctly, that because lunar gravity is weak, particles ejected by impacts on the back of the Moon might drift around and settle on the front. Thus, the sample from Mare Tranquilitatis might include samples from all over the lunar surface. Also, Wood was worried that the first mission to the lunar surface might be brief; the astronauts might simply open the door, grab a handful of dirt, and come home; soil might be all they brought back!

Wood had assembled a team of four: two meteorite experts, an electron microprobe technician, and me. Our objective was to characterize the lunar materials by describing their different mineral assemblages, textures, and chemical compositions. Like other investigators, we found that the soil was dominated by finely crystalline basalt and glass. There were many fused or fractured crystals—the product of shock metamorphism caused by billions of years of meteorite bombardment. There were also lots of dark glass beads, formed when impact melting threw droplets of silicate melt above the surface, which fell back as frozen raindrops. And then we found a few bits of rock that were white instead of black. These white rocks turned out to be 60 to 90 percent calcium-rich, plagioclase feldspar, with smaller amounts of olivine and/or pyroxene. Texturally, they did not look like lavas. They were more like shallow intrusive igneous rocks, which occur on Earth and are called *anorthosites*.

When we discovered the bits of anorthosite in the soil, we assumed that they had been formed by fractional crystallization because that is how, we believe, such rocks form in the Earth. I assumed that this process probably occurred in a lava lake on *Mare Tranquilitatis*. Given the facts at our disposal, it was a plausible, reasonable idea, and wrong. For some reason, John Wood did not accept my interpretation. He went home for the weekend and returned on Monday with a different idea—that the anorthosite came not from Mare Tranquilitatis but from the lunar highlands, Galileo's terrae, and that the fractional crystallization had occurred not in a lava lake but across the entire Moon. The Moon had melted!

I was appalled. Ockham's Razor seemed poised to strike.* How could we justify such a radical idea when simpler explanations were at hand? Why go to the highlands when we could make

---

*Named for English philosopher William Ockham (1285?–1347), although he did not invent it, Ockham's Razor, or the Principle of Parsimony, states, "It is futile to do with more what can be done with fewer."[100] In science simpler explanations are preferred to explanations that invoke needless complications.

anorthosites locally? Why melt the Moon when we could make enough anorthosite by fractionating a very small part of the mare basalt?

The strongest evidence in Wood's favor was correspondence between the composition of the anorthositic material in the soil and an analysis made by the *Surveyor VII* spacecraft, which landed on the blanket of material (ejecta blanket) thrown out from the highlands crater Tycho. Considering the dubious precision of *Surveyor VII*, however, and its inability to distinguish between calcium, phosphorous, sulfur, and potassium, or between iron, titanium, chromium, and nickel, I was not persuaded.

Element	***Anorthositic Fragment***	***Surveyor Analysis of Tycho Ejecta***
Carbon		< 2
Oxygen	60.9	58±5
Sodium	0.2	< 3
Magnesium	4.1	4±3
Aluminum	11.3	9±3
Silicon	16.2	18±4
"Calcium"	5.4	6±2
"Iron"	2.0	2±1

Note: Compositions are expressed as atomic percent.[101]

I fretted and fussed all the way to Houston, but my worries vanished at the First Lunar Science Conference when I learned what others had found. Our paper, incautiously titled "Lunar Highlands Anorthosite and Its Implications," was not scheduled until the end of the third day. At ten o'clock on the first morning, however, J. V. Smith, A. T. Anderson, and their colleagues from the University of Chicago reported that they had found anorthositic glass in the soil. They, too, had noticed the similarity to *Surveyor VII*'s analysis at Tycho, and they had the temerity to suggest that the glass represented impact melts from the highlands and that the Moon had undergone a major melting and crystal fractionation event! Wood was disappointed to be scooped, but I was relieved to be joined in our folly by such prestigious company. Then we learned two crucial facts. First, trace element analyses showed that the mare basalts contained lower concentrations than expected of the rare earth element europium. The experts suggested that europium had been extracted from the basalt, *or its source,* by fractional crystallization of plagioclase feldspar. Second, melting experiments showed that plagioclase was not an early (high-temperature) precipitate from the mare basalts. Crystal fractionation from these basalts would not produce plagioclase concentrates, which meant that the mare basalts were not the source of the anorthosites. All of a sudden, Wood's outrageous idea looked very, very plausible. Without a moment of hesitation, I hopped happily onto the bandwagon.

According to Wood's model, shortly after formation at least 50 percent of the Moon melted. Fractional crystallization of the magma ocean produced a floating crust of anorthosite, a sunken pyroxene-rich mantle, and, in between, a sandwich layer enriched in potassium, rare earth elements, and phosphorous (collectively referred to as KREEP). Impacts of large meteors created the mare basins, which were subsequently filled by magmas derived from the pyroxene-rich mantle. Later melting of the sandwich horizon yielded the rare KREEP basalts, rich in potassium, rare earth elements, and phosphorous.

In the 25 years since *Apollo XI,* many details have been added to and altered in our understanding of lunar history, but the basic idea remains. The Moon melted and underwent fractional crystallization on a giant scale. Twenty years before *Apollo XI* such melting had been suggested for Earth by F. A. Vening Meinesz and Harold Urey, introducing the idea of the core forming event, but that was just an idea. *Apollo XI* brought evidence that encouraged consideration of the possibility that Earth and the other terrestrial planets had also melted.

Moon rocks are very old by terrestrial standards. Few rocks on Earth are as old as the youngest lunar rocks, which formed 3.1 billion years ago. Shortly after that, the Moon's internal energy apparently ran out. The only subsequent changes to the Moon have been caused by meteorite impact and human beings.

The Moon had a magnetic field when its igneous rocks formed, and the rocks record that field as remanent magnetism.* The magnetic field apparently died when the Moon's internal energy supply

---

*That is, the magnetic crystals in the rocks point to the location of magnetic north as it was when the rocks were formed.

was exhausted. According to geophysicist S. K. Runcorn, the Moon's magnetic field was quite strong during the first few hundred million years of lunar history, which may mean that the young Moon had a small molten core.

Moon rocks are fundamentally similar to terrestrial rocks, but they show some differences. They contain no water, whereas every earth rock we can sample does. Also, other volatile materials, such as the alkali metals, are less abundant in the lunar crust than in the terrestrial crust. The oxidation state of lunar rocks is lower than that of Earth's crust, so, lunar rocks commonly contain native iron and nickel-iron alloys. Moreover, unlike terrestrial rocks, which generally have some amount of secondary alteration, lunar rocks are pristine except for the effects of cosmic rays and hypervelocity impacts. Moon rocks show no signs of dynamic metamorphism—there are no lunar schists or gneisses. Why? Because, aside from a few faults, there are no signs of tectonic activity on the Moon.

The Moon's density indicates that it contains much less iron than Earth. The density suggests that the lunar interior is dominated by silicates such as olivine and pyroxene. If an iron-rich core is present, it must be very small.

Where did the Moon come from originally? There are presently four hypotheses: Earth fission, intact capture, co-accretion, and mega-impact.

The Earth Fission Hypothesis. According to this idea, which was first proposed in 1880 by Charles Darwin's son George, the young Earth got to spinning so fast that it split apart, throwing out the material for the Moon. The resulting excavation in the Earth was the Pacific Ocean basin. This hypothesis has been stymied by the absence of a plausible mechanism to make Earth spin fast enough and by the observation that the total angular momentum in the Earth–Moon system is less than one third that required for fission.

The Intact Capture Hypothesis. The Moon formed elsewhere in the solar system, presumably farther from the Sun; was thrown in towards the Sun, as are the various Earth-crossing asteroids; and came close enough to Earth at the appropriate velocity to be captured in its orbit. This hypothesis seems possible but implausible. The dynamic conditions for capture are very special and therefore unlikely. Furthermore, the idea of lunar accretion at a different solar distance does not account for several chemical similarities between the Earth and the Moon. For example, the oxygen isotope ratios in the two are so similar that they suggest accretion in the same region of the nebula.

The Co-Accretion Hypothesis. Planetesimals (moon-sized objects) zooming around an accreting planet may collide and create a disk of particulate matter in orbit around the planet. Such matter might accrete into a moon. The co-accretion process almost certainly happens, but the hypothesis fails to answer several questions, such as: Why is the Moon so depleted in iron? Where did the heat come from to melt the Moon? How did the Moon acquire its observed angular momentum?

The Mega-Impact Hypothesis. One or more large planetesimals, perhaps as large as Mars, struck the Earth. The impactor and a portion of Earth's mantle were vaporized or melted, and much of the vapor and melt were captured in orbit around the Earth and subsequently formed the Moon. To date this hypothesis has uncertainties but no fatal flaws. The impactor must have been larger than the Moon, perhaps larger than Mars. (Mars' mass is 10 percent of Earth's and almost 10 times the Moon's.) Also, impact must have occurred after the Earth and the impactor were chemically differentiated. In order to keep iron out of the Moon, most of the iron in both bodies must have already been sequestered in central cores, and both Earth and impactor must have been hot. Under these conditions an impact would have produced an eruption of liquid and vapor away from the impact site, which would have carried part of the angular momentum of the impactor. From this ejected material, a disk would have formed, with liquid in the center surrounded by gas layers above and below. Except for hydrogen and helium, the disk would have retained most of this material. Thus, hydrogen would have been lost to interplanetary space, depriving the Moon of water. Within hours, the spinning disk would have split into materials that fell back into Earth (including, perhaps, the iron core of the impactor), and materials further out that coalesced to form "protomoonlets." These could have accreted to form the protomoon. If this happened quickly, the protomoon would have been mostly molten and susceptible to crystal fractionation. Like all serious ideas in science, this idea is subject to test, and the critical questions and answers may have as much to do with earth history as with lunar history.

Our knowledge of Mercury is based upon astronomical observations, hampered, unfortunately, by the glare of the Sun, and upon television images beamed back by the *Mariner* spacecraft in 1974 and 1975. Here's what we know:

Of all the planets only Pluto, the farthest from the Sun, is smaller and more eccentric in its orbit than Mercury, the closest. Mercury's orbit inclines 7 degrees to the plane of the ecliptic and ranges from 0.31 to 0.47 AU* from the Sun. It orbits the Sun in just under 88 days, spinning on its axis in the same direction, every 58½ days. This means that it rotates three times for every two solar orbits. Because its spin axis is perpendicular to the orbital plane, there are no seasons on Mercury, which has the most extreme temperature conditions in the solar system. Days are long (176 earth days) and hot enough to melt zinc (427°C); nights are equally long, and cold enough to freeze methane (–183°C).

Mercury is three times larger than the Moon and less than half the size of Mars. It is nearly as dense as Earth (5.4 versus 5.5 g/cm^3) because of its large, dense core, which is about the size of the Moon, three-quarters of the planet's diameter, and 42 percent of its volume. By comparison, the Earth's core is slightly more than half of the Earth's diameter and 16 percent of its volume. Another way to express the difference is to look at the uncompressed densities* of the two planets: Earth's uncompressed density is 4.0 g/cc^3; Mercury's is 5.3 g/cm^3. Like Earth, Mercury has a dipolar magnetic field aligned parallel to the spin axis, but the strength of its magnetic field is only 1 percent that of Earth's field. Mercury's core is probably solid iron.

The *Mariner 10* images of Mercury are Moon-like. Like the lunar surface, Mercury's surface has been shaped by impact cratering, but unlike the Moon Mercury is scribed by long arcuate scarps that

---

*Recall, from Chapter 1, that an astronomical unit, or AU, is the mean distance between Earth and the Sun, approximately 93 million miles.

*The uncompressed density is the density the material would have if it were at the surface of a planet rather than buried in the interior. Uncompressed densities make it easier to compare the internal compositions of planets of very different sizes.

appear to mark thrust faults, due perhaps to core solidification and planetary contraction.

A signal event in Mercury's history was a major impact that created the 1,300-kilometer-wide Caloris Basin. Shock waves from this impact may have been strong enough to create a distinctive hilly and lineated terrain on the opposite side of the planet. Caloris and other impact basins were flooded by volcanic materials whose composition—or that of the crust as a whole—is not known. We know that Mercury was differentiated into a core and overlying, less dense rock. The rocky portion is likely to be peridotite overlain by basalt.

Because Mercury's spin axis is perpendicular to the planet's orbital plane, the Sun is directly overhead along the equator and never shines into craters near the poles. Some of these high-latitude pits contain material with the radar-reflectivity of water ice. As *Sky & Telescope* magazine exclaimed, "Ski Mercury!"[102] Indeed, in spite of Mercury's proximity to the Sun, it appears that snow fields have lain in frigid polar shadows, perhaps for billions of years, since the cratering impacts that dug their niches. Someday we will sample these ancient ices.

Star light, star bright,
First star I see tonight,
Wish I may, wish I might,
Have the wish I wish tonight.

Venus is the brightest "star" in the sky, the evening or morning star upon which generations of men and women have pinned their hopes and dreams. Earth's twin and, perhaps, Earth's harbinger, she is an object of great scientific interest and wonder.

In 1610 Galileo first observed the phases of Venus and provided another powerful argument for the Copernican theory. Because Venus is shrouded in clouds, however, neither Galileo nor anyone else could observe the surface of the planet until after 1960, when radar signals and spacecraft could be used to penetrate the clouds. Since then we and the Russians have thrown a lot of hardware at Venus.

1960	First radar echoes confirm that Venus has a solid crust.
1961	*Venera 1* accomplishes the first flyby. Misses Venus by 100,000 kilometers.
1962	Radar doppler effects indicate that Venus rotates in a retrograde direction every 240 days.
1962	*Mariner 2* flyby. Misses Venus by 34,833 kilometers.
1965	*Venera 2* flyby. Misses Venus by 24,000 kilometers.
1965	*Venera 3* makes first hard landing on Venus.
1966	Areceibo, Goldstone, and Haystack antennas achieve radar ranging accuracies on Venus of better than 1 kilometer.
1967	*Venera 4* measures composition, pressure, temperature, and wind velocity in lower atmosphere.
1967	*Mariner 5* flyby. Measures density and composition of atmosphere.
1969	*Venera 5* and *6* atmospheric probes measure composition of atmosphere.
1970	*Venera 7* first soft landing on Venus. Direct measurement of surface temperature and pressure.
1972	*Venera 8* soft landing. Measures soil radioactivity.
1973	*Mariner 10* flyby. Investigates cloud dynamics.
1975	*Venera 9* and *10* soft landings. First television pictures of surface; soil radioactivity determinations.
1978	*Pioneer Venus I* spacecraft equipped with radar altimeter placed in orbit around

	Venus. Horizontal resolution, 30 kilometers, vertical resolution, 700 meters.
1978	*Pioneer Venus II*. Confirms absence of magnetic field.
1978	*Venera 11* and *12* flybys with landers. Atmospheric chemistry.
1981	*Venera 13* and *14* soft landings. Color television view of landing site; chemical composition of surface rocks.
1981	*Venera 15* and *16* orbiters. Radar mapping; horizontal resolution, 1–2 kilometers, vertical resolution, 30 meters.
1984	*Vega 1* and *2* flybys with landers. Atmospheric balloons deployed, detailed temperature profiles of lower atmosphere; chemical composition of soil.
1990	*Magellan* spacecraft placed in orbit around Venus; begins four-year program of radar mapping, altimetry, and gravity measurements.

In some respects, such as mass and overall density, Venus resembles Earth. In others, it is very different. Venus orbits the Sun every 225 days in a nearly circular orbit. It turns backwards on its axis, which is nearly perpendicular to its orbital plane, only once every 243 days. Consequently, the Sun rises in the west and sets in the east; days and nights are 58 Earth days long; and there are no seasons. The surface is even hotter (462°C) than the daytime surface of Mercury, and the atmosphere of Venus is about a hundred times more massive than Earth's.. Atmospheric pressure on the surface is 92 bars, which is slightly more than the pressure beneath half a mile of water.

The discovery of the atmosphere on Venus has a long history. Around 1660 Christian Huygens examined the planet with his telescope and, being unable to discern any surface features, suggested that the light from Venus was reflected by an atmosphere. One hundred years later when Venus made one of its infrequent transits across the face of the Sun, Russia's first great scientist, Mikhail Vasilyevich Lomonosov (1711–1765), observed some peculiar optical effects including momentary blurrings of the solar and planetary edges as Venus entered and left the solar disk. Lomonosov reasoned "that the planet Venus is surrounded by a considerable atmosphere equal to, if not greater than, that which envelops our earthly sphere."[103] He was right, but for the wrong reasons. The effects he observed were caused not by the atmosphere but by diffraction. A few years later, however, other astronomers did observe valid evidence of the atmosphere. Because of various spectroscopic difficulties—including interference by Earth's atmosphere—the principal chemical component of the venusian atmosphere was not identified for another 140 years, and the less abundant chemical components were not identified until space probes reached Venus in the late 1960s.

The atmosphere is dominated by carbon dioxide plus a whiff of nitrogen, and the clouds are sulphuric acid. A similar difference between atmosphere and clouds occurs on Earth, where, of course, the atmosphere is dominated by nitrogen plus 21 percent oxygen, and the clouds are water and ice. Earth is wet; Venus is dry. Earth contains 100,000 times more water than Venus (10^{24} versus 10^{19} grams).

The scorching heat at Venus' surface is the result of a runaway "greenhouse" phenomenon. Carbon dioxide (CO_2) is an infrared-active gas, which means that it is a strong absorber and emitter of infrared radiation. Sunlight penetrates Venus' cloud-filled atmosphere by a series of radiative absorptions and emissions, as well as by straight transmission, and heats the rocks below. The rocks glow with invisible, infrared waves, but these waves are not efficiently transmitted back to space. Most are absorbed by the atmosphere, which then emits infrared radiation upwards and downwards. Eventually the rocks on the surface get hot enough to emit enough higher-energy (shorter-wavelength) radiation, which can be transmitted through the atmosphere, to balance the incoming solar energy. For Venus this equilibrium is reached at 462°C.

Earth also maintains a radiative equilibrium. At the present time the composition and structure of our atmosphere acts to balance the incoming and outgoing radiation at an average surface temperature of about 22°C. The problem that will face our descendants, if it is not already upon us, is that various human activities are changing the balance by adding to the atmosphere significant amounts of CO_2 and other infrared-active gases which *may* eventually raise surface temperatures, alter ecosystems, and threaten the existence of many species, perhaps including our own. Uncertainty persists because the process will also add water to the terrestrial atmosphere. Water is an infrared-active substance—rain and snow clouds are strong absorbers and emitters of infrared radiation. However, such clouds also cool Earth by reflecting sunlight back into space. No one knows which effect will be stronger, the cooling effect due to reflection or the heating effect due to atmospheric absorption and emission. Venus may or may not be a terrifying harbinger of our fate, but whether Earth experiences global warming, another ice age, or perpetual gloom, our paradise will have been destroyed.

Some say the world will end in fire,
Some say in ice.
From what I've tasted of desire
I hold with those who favor fire.
But if it had to perish twice,
I think I know enough of hate
To say that for destruction ice
Is also great
And would suffice.

Robert Frost*

Another effect of the dense atmosphere on Venus is to moderate temperature changes. Nighttime temperatures are not much cooler than daytime temperatures, and the range from equator to the poles is only a few degrees. The absence of strong temperature differences, combined with the planet's very slow rate of rotation, results in static weather conditions at the surface.

Descending through thickening skies, the space probes encountered the sulphuric acid cloud deck 64 kilometers above the ground. At that altitude westward winds streamed by at more than 200 miles per hour. Twenty kilometers above the ground the probes broke out of the clouds and viewed a variegated volcanic terrain, in somber shades of umber and black, beneath a foul-smelling, dirty orange sky. Syrupy breezes of hot, dense gas swept slowly westward across the surface. The atmosphere was 50 times more dense than air on Earth's surface (63 g/L versus 1.2 g/L), extremely hot (462°C), and compressed (92 bars). Such temperature and pressure could be simulated at the bottom of a 270-foot-deep pool of molten lead, but the similarity ends there because molten lead is 150 times more dense than the atmosphere of Venus. The surface conditions of Venus can be created in a laboratory, but there is nothing remotely like them in normal human experience. Yet, in spite of these extraordinary conditions, the images sent back by the Venera landers were in many ways familiar. The sky was light, the landscape was gloomy but not dark, and the ground immediately around the landers was littered with broken slabs of rock, gravel, and sand.

The surface of Venus is covered by volcanic rocks. Unlike on Earth, where volcanoes are mostly restricted to plate boundaries or hot spots, on Venus they are everywhere. There are tens of thousands (perhaps millions) of small dome-shaped volcanoes, up to 20 kilometers across. Large volcanoes, with diameters greater than 50 kilometers, cover 5 percent of the planet's surface. Basaltic volcanism apparently predominates; some of the chemical analyses obtained by the Venera landers are basaltic, and most volcanic forms (cones, flows, and channels) suggest basaltic lava. Other chemical analyses indicate that more alkaline rocks are also present, and there are bulbous protrusions that resemble rhyolite domes on Earth.

Weathering and erosion have been much less important on Venus than on Earth or Mars and are, so far, apparent only in the effects of meteorite impact and landslides associated with tectonic uplifts or volcanism. Chemical weathering may be effective on Venus, but we have no evidence for it yet. Likewise, there may be erosion and deposition by the wind. Such effects have not yet been identified, although the steep slopes, rounded forms, and jagged surfaces seen in panoramic views created from satellite data could result from wind action. Water is not a factor in weathering or erosion. It has not flowed on Venus for at least a billion years, if ever.

Because most meteors burn up in the thick atmosphere, there are only about 900 impact craters on the surface. These craters are randomly distributed and suggest that the surface of the planet is everywhere about the same age, which is estimated to be 300 million to 600 million years. Large volcanoes have fewer impact scars than the global average. A likely interpretation of this is that a few large volcanoes continued to erupt after tectonism and most volcanism had already ceased.

From the days of Alfred Wegener we have known that topography on Earth is bimodal: 35 percent of the surface is elevated continental crust, and 65 percent is depressed ocean basin. On Venus topography is unimodal: There are definable highlands and lowlands, but 80 percent of the surface lies within 1 kilometer of the average elevation. This difference in topography is a manifestation of fundamentally different tectonic and erosional processes at work. Earth's physical, chemical, and biological nature, from the bottom of the core to the top of the atmosphere, is affected by global plate tectonics, the results of which are the aggregation of great continental rafts of low-density rock and the perpetuation of deep ocean basins. Without the efforts of this great tectonic engine the topographic range of the planet would be reduced and simplified, as it is on Venus.

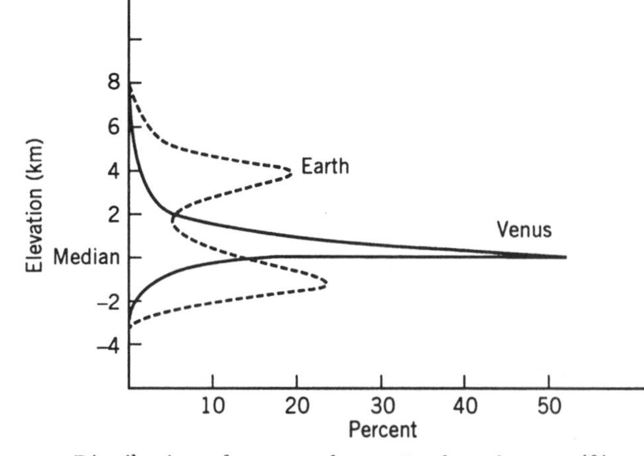

Distribution of topography on Earth and Venus.[104]

Venus is divided into three topographic domains. Forty percent of the surface lies below the average elevation; these lowlands have repeatedly experienced volcanic and tectonic activity over wide areas. Twenty percent of the surface lies more than 2 kilometers above the average elevation; these highlands include two regions that resemble small continents, both roughly elliptical and elongated along their east–west axes. *Ishtar Terra,* which stretches one third of the way around the globe between 60 and 90 degrees north latitude, is about the size (but not the shape) of Australia. The highest mountain of Ishtar Terra, Maxwell Montes, rises 12 kilometers above the lowlands and is the highest mountain on Venus. Aphrodite Terra extends halfway around Venus along the equator, is about the size (but, again, not the shape) of Africa, and rises about 3 kilometers above the lowlands. Western Aphrodite is characterized by compressional features, but eastern Aphrodite has extensional rift valleys and volcanoes. The remaining 40 percent of the planet's surface is covered with rolling plains of intermediate elevation.

Tectonic features on Venus include linear mountain belts with ridge and valley topography reminiscent of the Appalachians. Such features are unique to Venus and Earth. Rift zones up to 300 kilometers wide and 5 kilometers deep extend for thousands of kilometers. Extensional features like these are common on Earth and Mars as well. Unique to Venus are large circular features called *coronae,* which are the surface expressions of some unexplained mantle upwelling. Coronae are usually a few hundred kilometers across, but some are as wide as 2,600 kilometers. They may have radial or concentric fracture systems and are often surrounded by arcuate ridges and depressions. Other curious features, which look like spiders, pancakes, or piles of shattered tiles, are common manifestations of volcanism and/or tectonism.

Tectonism, like volcanism, is ubiquitous on Venus, but the tectonic elements, folds and faults, rifts and ridges, are not articulated in a global tectonic system. There is no evidence of plate tectonics. Some extensional areas may correspond to accreting plate margins on Earth, but nothing has been seen on the scale of Earth's mid-ocean ridge system, and there are no great faults like terrestrial transform fault plate boundaries. Great faults, like those that apparently cut completely through Earth's lithosphere from top to bottom, may be impossible on Venus because its lithosphere may be too dry, too thick, and too strong. Covered as it is by a burning hot surface, the lithosphere of Venus must be hundreds of degrees hotter than Earth's. Subduction (sinking) into Venus may be impossible because the lithosphere is too hot and therefore too buoyant.

Venus is presumed to be differentiated, like Earth, into a crust, a mantle, and a core. Because no one has yet landed a seismometer on the planet, however, we have no seismic data to reveal the interior structure. Instead, models of the interior are based upon studies of deep impact craters, tectonic structures, and the relationship between changes in gravity and topography. The following interpretations are thought to be robust: First, the normal crust is about 20 kilometers thick, which makes it more like Earth's oceanic crust than its continental crust. High areas, such as Ishtar Terra and western Aphrodite, may have crust as thick as 40 kilometers, making them more like terrestrial continents. Second, the lithosphere of Venus is at least 70 kilometers thick and strong enough to have supported many topographic features for hundreds of millions of years. The strength of the lithosphere, as suggested above, may derive from its extreme dryness. Third, unlike Earth, Venus has no low-velocity–low-viscosity zone beneath the lithosphere. Its lithosphere lies directly upon the underlying, convecting mantle.

What of the deeper mantle and core? The deep mantle of Venus is thought to behave like the deep mantle of Earth, convecting heat slowly up from the core. The core of Venus does not behave like Earth's core, however. Unlike in Earth, convection in the core of Venus does not generate a strong magnetic field. Why? Perhaps it lacks sufficient energy to drive convection cells fast enough. Why? Perhaps because no inner core is forming and releasing its latent heat of crystallization. This source of heat is thought to be important in Earth's core. Or perhaps, as physicist James A. Van Allen suggests, the weak magnetic field results from the planet's exceedingly slow rotation on its axis.[105] Why is Venus turning so slowly, and why is it turning backwards? We don't know. Perhaps the planet was hit by a planetesimal going the "wrong" way. Perhaps, perhaps . . .

The information sent back by orbiters and landers and the information provided by Earth-based radar experiments has vastly increased our knowledge of Venus, but our understanding of how the planet evolved and operates is still very incomplete. The topic is one of active inquiry, uncertainty, and debate that exemplifies science in progress. Consider the last sentence of a recent offering on the subject: "As with previous attempts to describe Venus, parts of this model will be wrong, but hopefully other parts can be carried forward."[106] *Sic transit scientia!*

Mars is much smaller than the Earth or Venus. It's mass is 10 percent of Earth's; its radius is half of Earth's; and its density is only 3.95 g/cm^3 compared to Earth's 5.52 g/cm^3. Mars is much colder, with surface temperatures ranging from 0° to minus 125°C. On the other hand, it is only a few months' space flight away; it has water; it has sunshine; and it has a bright sky. Before the twenty-first century is half over, people will walk on Mars. For our great-grandchildren, spending a field season on Mars will be an adventure comparable to today's excursions to the Antarctic.

The orbit of Mars is significantly more eccentric than those of Venus, Jupiter, or Saturn. Johannes Kepler discovered the elliptical motion of the planets by choosing to study Mars. Given the imprecision of pretelescopic astronomy, it was a fortuitous choice because, as he wrote, "to arrive at the secret knowledge of astronomy, it is absolutely necessary to use the motion of Mars; otherwise it would remain eternally hidden."[107]

Because of its eccentricity, Mars approaches Earth more closely than usual every 15 years. One such close approach occurred in 1877, the year that Giovanni Schiaparelli first saw *canali,* the imaginary canals that led Percival Lowell and millions thereafter into a dream world of little green men. Earth-bound telescopes have identified real features as well: In the seventeenth century, Christian Huygens first saw the dark triangular region, known as *Syrtis Major,* and the south polar ice cap. A few years later, Giovanni Cassini, discoverer of Saturn's rings, determined that the Martian day was a little more than twenty-four hours long. In the eighteenth century, William Herschel suggested that the polar caps were water ice. He also discovered the Martian atmosphere and measured the inclination of the planet's axis of rotation. In the nineteenth century, astronomers identified Mars' two moons, Phobos and Deimos, as well as transient white clouds and yellow clouds. They also described, but did not identify, the great volcanoes Olympus Mons, Arsia Mons, and Ascreus Mons.

The modern era of Mars exploration began on July 16, 1965, when *Mariner 4* passed within 9,780 kilometers of the red planet. The pictures *Mariner 4* sent back showed only a few craters on a featureless plain. By chance, the first flyby had scanned the least interesting part of Mars. Subsequent missions, including soft landings, have revealed a diverse, geologically interesting planet.

Unlike the exploration of Venus, the United States has been responsible for all of the significant achievements at Mars. The Soviet attempts were generally unsuccessful.

1965	Mariner 4 accomplishes the first flyby. Takes 22 pictures with 3-kilometers resolution; measures atmospheric density and magnetic field.
1969	*Mariner 6* and *7* flybys take 58 pictures; identify atmospheric molecules; measure surface temperature.
1971	*Mariner 9* orbiter photo maps planet at 1-kilometer resolution; takes pressure-temperature and composition measurements of atmosphere. Discovers numerous geologic features, including giant volcanoes and canyons. Soviet *Mars 2* and *3* orbiter/landers. No useful data returned.
1974	Soviet *Mars 4* and *5* orbiters, *6* and *7* landers. Orbiters obtain more pictures, atmospheric composition; landers unsuccessful.
1976	*Viking 1* and *2* orbiter/landers. High-resolution imagery, gravity measurements; landers test for life, analyze soil, monitor seismic activity, take pictures of landing site.

The atmosphere of Mars is very thin, with an average pressure at the surface of 0.008 bars. Its composition is dominated by carbon dioxide, plus less than 3 percent nitrogen. The white clouds are snow clouds. Like Earth, Mars has an ozone (O_3) layer caused by solar radiation, but because Mars' atmosphere is so much thinner than the Earth's, the Martian ozone lies close to the ground. Ozone is a powerful oxidizing agent and may be the principal reason for the highly oxidized state of Martian soil and the planet's red color. The color is, of course, Mars' distinctive characteristic, the inspiration for its name and menacing reputation. Astronomers with telescopes early on noticed light and dark regions on the surface. These are real and due to the shifting distribution of yellow, windborne dust on darker bedrock.

Æolian (windborne) deposits cover most of Mars, so our knowledge of crustal bedrock variations is significantly constrained. Although surface color and reflectance variations help us understand the Moon or Mercury, they tell us little about the composition of the ancient Martian crust, for most of it is covered by dust and sand. Mars is subject to annual, planet-wide dust storms. Such a storm was in progress when *Viking 1* arrived there in 1976, so the first pictures of the surface show only Olympus Mons poking up through swirling clouds of mustard-yellow dust.

What do we know of the bedrock? Color images beamed back by *Viking Landers 1* and *2* show landscapes strewn with dark boulders that look like basalt or some other dark, igneous rock. The Vikings provided chemical analyses only of the soils, which were similar at both landing sites and were composed mostly of iron-rich silicates. There were some surprises: Sulfur was two orders of magnitude more abundant than in Earth's crust; the concentration of potassium was less than one-fifth that of average crust on Earth; and aluminum was present in low concentrations. These soils, which do not necessarily represent average Martian crust, probably contain iron-rich clays, sulfates, carbonates, iron oxides, quartz, and, perhaps halite. This suggests weathering of rocks, such as basalt, in a highly oxidizing, arid environment. Spectral data collected by the *Phobos II* spacecraft, which orbited Mars in 1989, also indicate the presence of basaltic rocks on the Martian surface.[108]

The Martian surface has an ancient, cratered highlands terrain, upon which younger geologic features and contemporary æolian features have been superimposed. We do not know the composition of this ancient crust, which dates back 4 billion years and dominates the southern hemisphere and parts of the northern hemisphere. These surfaces have been intensely bombarded, like the ancient surfaces of the Moon and Mercury, but unlike on those airless places, erosive and depositional processes on Mars have smoothed the cratered highlands so they look less battered. There is, however, an even more dramatic difference: Many of the Martian impacts look like splatter marks in wet mud! Ejecta blankets spreading out from some impact centers look like very wet mud slops. What happened? We don't know, but we do know that Mars has been awash in water at times in the past.

Incised upon the highlands are outflow and runoff channels carved by flowing water. These channels were apparently fed by colossal outpourings of water early in the history of the planet, perhaps 4 billion years ago. The flows appear to have been singular events, as there is no evidence of an ongoing hydrologic cycle. There are no rainfed watersheds or persistent catchment basins such as lakes or seas. The waters appeared, flowed, and disappeared. Where did they come from? Apparently from the ground. Where did they go? We don't know. Back underground? Perhaps. There may be a lot of ice in the ground. A great deal of water may have evaporated and dissociated in the Martian atmosphere, and if that occurred, Mars would have retained the oxygen but lost much of the hydrogen to space. The development of the flow channels indicates that, unlike today, water was once stable on the Martian surface. Atmospheric temperatures and pressures were apparently higher than they are today. If so, Mars may once have been hospitable to life.

Today, in addition to the ice presumed to exist in the ground, Mars has water in the polar ice caps and in clay minerals of the soil. The ice caps wax and wane with the seasons. Frost appears in mid-latitude deserts in winter. Also in winter, deposits of dry ice (carbon dioxide) coat the perennial ice sheets. Beneath these sheets are layered deposits of ice and sediment, which may record climate changes early in Martian history.

Unlike the Moon and Mercury, Mars did not stop evolving billions of years ago. Most of the northern hemisphere is younger, lower, and shows the effects of volcanism, tectonism, and surficial processes involving wind, water, and ice. Giant volcanoes and a canyon complex are Mars' most spectacular, younger geologic features. Of the volcanoes, Olympus Mons is the largest, more than 700 kilometers across at the base and 25 kilometers high. It is the largest volcano in the solar system—nearly six times wider at the base and three times higher than Mauna Loa, the tallest (though not the highest) mountain on Earth. Olympus is located on the northwestern edge of a huge welt called the Tharsis bulge. Running atop the Tharsis bulge, probably following a fracture, are three more volcanoes that reach the same elevation as Olympus. These and Olympus are the four largest volcanoes on Mars, but there are many more that dwarf their terrestrial cousins.

How did such a small planet make such big volcanoes? Perhaps Mars had powerful thermal plumes but no plate tectonics. Unlike terrestrial chains of volcanic islands and seamounts (submerged volcanoes), which seem to have passed across localized magma sources, the Martian volcanoes may have sat still and received greater quantities of lava. There are more than 50 volcanoes in the Hawaiian Islands–Emperor Seamount chain. If all that lava had been piled up in one place, it would have made a volcano as big as Olympus Mons.

Crater counts and stratigraphic relationships indicate that Mars experienced a billion and a half or more years of volcanism, during which broad plateaus were built in the northern hemisphere. Following this early stage, for the past 2½ billion years, most volcanism has occurred around the Tharsis bulge, where the youngest volcanic activity may be quite recent, certainly not more than a few hundred million years ago. Dinosaurs with telescopes, or their mammalian successors, could have watched Olympus Mons erupt.

The canyon complex that dominates the younger Martian landscape is called Valles Marineris. It appears as an ugly gash across the planet's midriff, more than 4,000 kilometers long, more than 700 kilometers wide, and 7 kilometers deep. Actually, Valles Marineris is composed of interconnected canyons, which individually may be as wide as 200 kilometers. Compare this colossal chasm with the Grand Canyon of Arizona, which is 450 kilometers long, generally less than 30 kilometers wide, and 2 kilometers deep.

When we sent the Viking landers to Mars in 1976, the great question on everyone's mind was whether or not life exists there.

The answer is almost surely no, given that the Viking landers found no life and no organic compounds. The absence of organic material in the soils argues strongly against life on the planet today, but it does not argue against life in the past because organic compounds would quickly disappear in the oxidizing environment. Our great grandchildren may find fossils on Mars.

Mars has two very small moons, Phobos and Deimos, which were discovered in 1877. One hundred years later, the *Viking 1* orbiter sent close-up pictures of them back to Earth. Phobos, the largest, is only about 20 kilometers in diameter and looks like a chunky potato. Deimos, roughly half the size of Phobos, is smoother but also potato-shaped. Both are cratered, dark, and weakly reflective. A likely guess is that they are carbonaceous materials captured from the asteroid belt.

In our efforts to compare, contrast, and eventually understand the internal mechanics of Earth and the other rocky planets, much has been learned by considering the extent to which topography and gravity are correlated. We do this by superimposing topographic maps upon so-called geopotential maps, or *geoids*.

A convenient way to describe planetary variations in gravity is to draw a surface of equal gravitational potential centered upon some specified height above the center of gravity. This equipotential surface, called a geoid, undulates above the center of gravity, rising and falling with gravity variations so that lifting an object from the center of gravity to any point on the surface requires the same amount of work. For Earth the specified elevation is usually taken as mean sea level. The surface of the ocean corresponds closely to a geoid. If there were no currents, tides or, waves water would accumulate here and flow away from there until every spot on the surface of the ocean had the same gravitational potential or, in other words, until there would be no further tendency for any drop of water to flow sideways.

Any surface of still water is an equipotential surface. If we wanted a physical model of the geoid for the entire planet we could create one by cutting canals through the continents from ocean to ocean and observing the equilibrium water levels. It is easier, however, to determine the geoids of Earth and other planets by observing satellite motions.

A satellite in orbit is a balancing act between the gravitational force that draws it down and the tangential force that throws it out. If Earth's mass were distributed uniformly, the gravitational force between Earth and a satellite would be constant, and the satellite would travel along a smooth elliptical path with Earth at one of the foci. Earth is not so uniformly constructed, however, and satellites respond to irregularities in the gravitational field by departing from ideal elliptical orbits. If the gravitational force increases, perhaps because of a local mass excess in the planet, the satellite is drawn down closer to the planet's surface and speeds up. Conversely, if the force decreases, the satellite flies higher and more slowly. These variations have been observed in satellites orbiting the Moon, Venus, Mars, and Earth, and they have been used to describe variations of the gravitational fields associated with those bodies.

The relationships between the geoid and topography provide clues about a planet's interior. For Venus and Mars the geoidal and topographic surfaces correlate closely. Topographic highs, such as volcanic piles, correspond precisely to regions of increased gravitational potential. Topographic surfaces poke out where increased gravity would be expected to draw them in, implying either that the topographic features are so young that the planets have not had time to equilibrate, that the elevated masses are held up by the strength of what lies below (as a paper weight is supported by a table), or that the masses are held up by internal forces related to active thermal convection. On Venus, current interpretations of the data suggest that most features are supported by a strong, rigid lithosphere; in addition to lithospheric stiffness, the prominent highlands Ishtar Terra and Aphrodite Terra may also be supported by low-density roots, and one volcanic pile, called Beta Regio, may be supported by a thermal plume. On Mars, the elevated masses of Olympus Mons and the Tharsis volcanoes appear to be supported by a cold, thick, strong lithosphere.

On Earth the geoid correlates much less closely with topography. Unlike the geoids of Venus and Mars, ours does not indicate the positions of great mountain ranges or even continents. Large portions of the mid-ocean ridge and rise system correlate with gravitational highs, indicating excess mass accumulations at the surface.

Some regions that overlie major subduction zones (e.g., parts of the southwest Pacific area), also correlate with gravity highs, indicating mass accumulations at depth presumably related to sinking dense, lithospheric slabs. Hawaii is located on a strong positive feature, as is Iceland, and both are attributed to mass accumulations at the surface. Yet no other volcanoes correlate specifically with strong geoidal highs. The general lack of correlation between the terrestrial geoid and topography reflects the high degree of isostatic compensation in Earth.*

On the Moon circular gravity highs are associated with six circular maria. These "mascons" (mass concentrations) are due to accumulations of lava in the impact basins and also, perhaps, to post-impact intrusion of dense mantle rocks. They have survived for nearly 4 billion years because the Moon has been cold and rigid for most of that time.

*Recall from Chapter 6 that isostatic equilibrium is a condition in which mountains of lower density crustal rocks float in the higher density mantle, with shallow or deep roots according to their heights. As they erode with age the mountains displace less mantle material and float upward as small boats rise when passengers disembark.

Twelve

Worlds More Distant and Strange

Between the orbits of Mars and Jupiter the solar system changes from rocky planets of silicates, oxides, and iron, to planets and moons dominated by the lighter elements, hydrogen, helium, carbon, nitrogen, and oxygen. Presumably, this change was due to lower temperatures and pressures within the condensing nebula. Outward from the newborn Sun, somewhere between Mars and Jupiter, the temperature and pressure crossed the "frost line." Water ice became stable; hailstones formed on nuclei of silicate and oxide dust; and snow began to fall. The greatest mass of material accreted to form Jupiter, which is the planet most like the Sun in composition. Saturn, Jupiter's neighbor to the outside, also gathered a great mass of rocks, ice, and gas. Jupiter's neighbor to the inside never formed or was destroyed. Although predicted by the Titius-Bode Series, discussed in Chapter 1, there is no planet between Mars and Jupiter, only the fragments of one or more broken planets and planetesimals. Occasionally these fragments are dragged from their orbit by Jupiter's mighty gravity and fall into the inner solar system. We known them as *meteorites*.

In ancient Egypt, the word for iron derived from a phrase meaning "metal from heaven." Ancient scholars generally accepted the notion of heavenly objects falling to Earth, but this knowledge disappeared from Europe during the Dark Ages and did not reappear until late in the eighteenth century, when a German lawyer and physicist named E.F.F. Chladni proposed that masses of iron, discovered in Siberia and elsewhere, were spent fireballs that had fallen to Earth. As Chladni's proposition became accepted, meteorites became objects of great interest. Within the next hundred years, the basic chemical and mineralogical characteristics of common meteorites were described, leading to speculations about Earth's origin and interior. Sixty years before Oldham discovered the core, Gabriel August Draubrée (1814–1896), a French geologist and mining engineer, suggested that the Earth has a core comparable to the iron meteorites. Today, our ideas about the Earth's interior and origin are strongly influenced by the chemistries and mineralogies of meteorites.

Meteorites are the oldest pieces of matter we have access to; some go back to the very beginning of the solar system, over four and a half billion years ago. Because their orbits are observed to be elliptical, and at a low angle to the plane of the ecliptic, meteors are believed to be from the solar system. We can guess at the sources of meteorites, but we know the orbital trajectories for only a few. Pibram, Czechoslovakia was the site of a meteorite fall in 1959. By great good fortune the meteor's flight was observed by several meteor cameras, and the trajectory established by the photographs showed that it came from the asteroid belt, between the orbits of Mars and Jupiter. Other meteors, which did not result in meteorite falls, have been photographed passing through the Earth's atmosphere. Their trajectories also suggest orbits that reach aphelion* in the asteroid belt. For this reason, the asteroid belt is the presumed source of most meteorites, but some may have originated on the Moon or on Mars.

As we saw in Chapter 1, the relationship known as the Titius-Bode Law predicted a planet between Mars and Jupiter, about 2.8 astronomical units from the Sun. On the first evening of the nineteenth century, a monk in Palermo, Sicily, Giussepe Piazzi (1746–1826), discovered a faint, slowly moving object just where the Titius-Bode Law predicted. He named it Ceres, in honor of Sicily. Nine hundred and forty kilometers in diameter, Ceres is the largest of the asteroids and contains one-third of the entire mass of

---

*Aphelion is the farthest point from the Sun.

asteroidal material. More than 5,000 additional asteroidal objects have been discovered since Piazzi and most of them, the Main Belt asteroids, are in solar orbit between Jupiter and Mars. A few cross the orbit of Mars and come as far in as Earth's orbit. Fewer still cross within Earth's orbit.

In recent years Eugene and Carolyn Shoemaker of the U.S. Geological Survey have been systematically searching the sky for asteroids and comets that cross within the orbit of Earth. In 1990 they had identified 55 Earth-crossing asteroids in the 0.9- to 1.7-kilometer size range, and they estimate that there might be as many as a thousand.*

Some asteroids pass quite close to the Earth. In 1932 Apollo, which is 700 meters in diameter, came within 11 million kilometers of Earth; in 1936 and 1937, Adonis and Hermes, each about 500 meters in diameter, came within 2 million and 800,000 kilometers, respectively. The closest known flyby occurred in March 1989 when a previously unknown asteroid, 200 meters in diameter, came within 690,000 kilometers. To place this in perspective, the Moon is a little more than 380,000 kilometers from Earth. Smaller objects come closer still. In 1972 a million-ton object, 80 meters across, flashed through the atmosphere over North America. And, of course, some of these rocks hit the ground.

By studying the geologic record the Shoemakers have estimated the production of impact craters on Earth over the past 100 million years. Their data, arranged according to the minimum crater diameter, are as follows:[110]

Minimum Crater Diameter (km)	*Number of Impacts**
10	1,090
20	240
30	97
50	18
60	10
100	2
150	1

*Twenty-five percent of the small (less than 40-kilometers) impacts are attributed to comets, 75 percent to asteroids. Most of the larger impacts are attributed to comets.

---

*1,080 ± 500.[109]

Strong evidence indicates that major impact events have caused specific mass extinctions in Earth's history. One, the Cretaceous-Tertiary mass extinction, which marked the end of the Mesozoic Era and finished off the dinosaurs, almost certainly resulted from an asteroidal impact. Of the relationship between evolution and impact events the Shoemakers write as follows:

> The essential point is that evidence accumulated during the 1980s suggests that the collisions of large objects with the Earth have played a major role in the destruction (and evolution) of life here. When an impact triggers the loss of species and even whole families of organisms, ecological or environmental space opens up for new ones. Various species of mammals, for example, multiplied rapidly after the Cretaceous-Tertiary extinction. It can be argued that the presence of the human race on Earth may be due, in no small part, to chains of events initiated by large impacts abut 66 and 35 million years ago.[111]

When might we expect our next catastrophic impact? No one knows, but the Shoemakers have discovered a likely candidate. Amun, discovered by them in 1986, is 2 kilometers in diameter and composed of metal, undoubtedly nickel-iron. Its orbit suggests that Amun will probably hit Earth sometime in the next 100 million years.

What might a catastrophic impact be like? Who knows? Perhaps the Cretaceous-Tertiary event was as I have imagined:

> Shortly after dawn, ochre and grey cirrostratus clouds swept in from the west. By noon the sky was completely overcast and a light rain was falling. Thus began ten thousand years of terrestrial misery. Earth had encountered a swarm of comets or asteroids that bombarded her repeatedly with explosive, dust-raising impacts. Snow, sleet, and rain fell for months on end, often grime-laden, foul-smelling rain that made even the thick skin of an ankylosaur itch. From time to time the overcast sky would lighten, even clear, but then clouds would roll in and block the sun once more. Without sunshine, food chains collapsed.

For dinosaurs, bewildered, cold, sullen, and silent, in rocky caves or ponds and rivers, the situation was soon critical. Within a year half of them were dead from hypothermia, starvation, or disease; all were gone by the end of the decade. And it wasn't just the great reptiles who perished: Before the climate had fully recovered, 60 percent of all species living on the face of the Earth had been wiped out.

Earth is a small planet, and life, Earth's most fragile feature, endures in trembling balance with a climate that can be perturbed by massive atmospheric additions of dust, soot, or noxious chemicals. Millenia later a far different world awoke to blue skies and sunshine, a world of refugees who had survived by wit, courage, and luck. A world of small mammals, seed-bearing plants, and turtles.

* * *

From earthbound observations of reflected light spectra we have learned that there are several kinds of asteroids. Three-quarters of them are dark, as if covered by coal dust, the rest reflect as silicate rocks, metals, or mixtures of rock and metal. These various types are distributed systematically in the Main Belt, with the darker variety increasing in abundance outward from the Sun.

In October 1991, the *Galileo* spacecraft passed within 5,300 kilometers of an asteroid named Gaspra and recorded a video image of it. Gaspra is a potato-shaped, gray object, 19 kilometers long and not quite 9 kilometers across. The image from *Galileo,* with resolution down to 55 meters, shows that it has hundreds of impact craters. All in all, Gaspra looks a lot like Phobos and Deimos, the moons of Mars.

Studies of meteorites show that they come from several parent bodies: those that have melted and undergone igneous differentiation and those that have not. The former include pieces of nickel-iron alloys ("irons"), mixtures of nickel-iron alloys and silicate materials ("stony-irons"), and rocks with igneous textures ("achondrites"). The Achondrites are so-named because they lack the definitive characteristic of the other great meteorite source: *chondrules,* which are tiny spherical masses of crystalline matter (usually olivine or pyroxene) that apparently quenched from droplets of liquid. Meteorites that contain chondrules are called *chondrites* or *chondritic meteorites*.

Achondrites, stony-irons, and irons come from several different parent bodies, which were differentiated into metallic cores overlain by silicate mantles. Most of these are thought to have been in the asteroid belt, but a few seem to have come from the Moon or even Mars. Chondrites are very primitive condensates of the solar nebula that never participated in the igneous differentiation of a planet. They have, however, been part of accreted planetesimals and have been more or less thermally metamorphosed.

Chondrites contain silicates of iron, magnesium, calcium, and aluminum, such as olivine, pyroxenes, and plagioclase feldspar. They also contain abundant sulfides and nickel-iron alloys. In terms of trace elements the chondrites more closely resemble the Sun than does Earth, and they are very similar to one another. The principal variations between them are in degree of oxidation, sulfur content, and carbon and hydrogen content. Those with the most carbon are called *carbonaceous chondrites*.

> "Well," said Owl, "the Spotted or Herbaceous Backson is just a—."[112]
>
> *A. A. Milne*

Carbonaceous chondrites represent the least altered parts of that region of the solar nebula from which the terrestrial planets are thought to have formed. In other words, if you licked the bowl from which the Earth was mixed, it would taste rather like carbonaceous chondrite meteorites. Although these meteorites contain individual crystals that formed at much higher temperatures, we know that after the meteorites accreted they were never hotter than a warm summer day (27°C).

Compared to other chondritic meteorites, carbonaceous chondrites contain more free carbon and usually more water, hydrocarbon compounds, and sulfur. They are apparently quite common in the solar system—the dark objects that dominate the asteroid belt are assumed to be carbonaceous chondrites, as are the moons of Mars.

All carbonaceous chondrities are primitive materials, but in 1969 Earth was visited by one of the oldest of all. In the predawn of

February 8, a carbonaceous chondrite broke up over the Pacific Ocean, and more than two tons of its dismembered body fell upon a tiny hamlet in central Mexico, Pueblito de Allende. It came to be called *Allende*.

Up in Cambridge, Massachusetts, John Wood's team was preparing for *Apollo XI*. Because we were associated with the Smithsonian Institution we were privileged to receive one of the first pieces of Allende to study, a fragment about the size and shape of a football. We treated it as a special opportunity to test our readiness for *Apollo XI*.

For several weeks we analyzed and described what we saw in the Allende meteorite. I had never described a meteorite before, so for me it was all exciting. For the experienced meteorite scholars the presence of glass and certain aluminum-rich minerals, such as spinel ($MgAl_2O_4$), was particularly interesting; we built our report around those observations.

When the analytical work was complete, and the report mostly written, John Wood went off to Houston for yet another pre-Apollo meeting. I was sitting at my desk one afternoon when Ursula Marvin walked in. She was a recent Ph.D. from Harvard, but had been a member of the Smithsonian scientific staff for nearly a decade and was already an international authority in meteoritics.

"Have a look at this," she said.

What Ursula then showed me was a list of chemical compounds that physical chemist Harry Lord III had calculated to be the compounds that would condense at the highest temperatures from the cooling solar nebula. She had remembered Lord's work and unearthed it from her files. The correspondence between Lord's list and the minerals we had found in Allende was suggestive. Of Lord's dozen chemical compounds we had found four in Allende, and a couple of others looked like possible substitutes.

High Temperature Condensates Predicted by Harry Lord	***Corresponding Minerals Found in Allende by Marvin et al.***
$MgAl_2O_4$	Spinel, $MgAl_2O_4$
$CaTiO_3$	Perovskite, $CaTiO_3$
$CaAl_2Si_2O_8$	Anorthite, $CaAl_2Si_2O_8$
$CaMgSi_2O_6$	Diopside, $CaMgSi_2O_6$

One thing was clear: the paper had to be rewritten, and there was another change to be made. I called Wood in Houston:

"John, Ursula has found a remarkable thing . . . Oh, and John, we need to change the order of authorship."

So the paper was published with Ursula Marvin as first author. She had discovered that the carbonaceous chondrite that fell upon Pueblito de Allende was the most primitive nebular condensate ever found. This was an important discovery and another example of Pasteur's dictum that "Chance favors the prepared mind." Allende went on to become the most intensively studied of all meteorites and a key to the solar system puzzle.

* * *

Jupiter is the largest planet, with a mass more than 300 times that of Earth yet only one-thousandth that of the Sun. Ninety-six percent of the planet is gas; the rest, way down inside is a small core, probably composed of rocks and ice. Jupiter has a powerful magnetic field whose interactions with the solar wind generate radio waves that were first detected in 1955. The planet emits twice as much heat as it receives from the Sun, probably because the gaseous giant is still collapsing. Images of the atmosphere show great bands of red, brown, and white materials streaming along latitudinal tracks, and within these bands, plumes, whorls, and clouds appear and disappear. The Great Red Spot is a large cyclonic disturbance that has persisted for at least three centuries, rolling counterclockwise between westward and eastward currents.

Jupiter has 13 satellites. The four largest, Io, Europa, Ganymede, and Callisto, which are the size of small planets, were discovered by Galileo:

> Accordingly, on the seventh day of January of the present year 1610, at the first hour of night, when I inspected the celestial constellations through a spyglass, Jupiter presented himself. And since I had prepared for myself a superlative instrument, I saw (which earlier had not happened because of the weakness of the other instruments) that three little stars were positioned near him—small but yet very bright. . . . On the thirteenth, for the first time

> four little stars were seen by me in this formation with
> respect to Jupiter . . . [113]

Galileo proposed to name them the Medicean Stars in honor of his patron, Cosimo II de' Medici, but history, more appropriately, knows them as the Galilean satellites. Their discovery accelerated acceptance of the Copernican system of planets in orbit around the Sun. And their regular risings and settings provided common time signals all over the Earth for eighteenth- and nineteenth century geodesists.

The Galilean satellites have nearly circular orbits, which lie in the plane of Jupiter's solar orbit. Closest to Jupiter is Io, the "pizza moon," which is covered with orange, brown, and yellow deposits of sulfur and sulfur compounds, especially sulfur dioxide. Io is the site of frequent volcanic eruptions, powered apparently by tidal friction. In its weak gravity which is comparable to that of the Moon, volcanic eruptions shoot hundreds of kilometers above the surface and fall back in umbrella-like cascades. Upon the surface are many vents, but no large volcanoes. Flows are long, indicating lavas of low viscosity. There are no impact craters on Io. Beneath the sulfurous crust lies molten sulfur and beneath that an interior of silicate melt and rock.

Next out is bright Europa, similar to Io in size but composed of a shining ice crust on an interior of silicate rock. Europa's heavily fractured but crater-free surface is apparently renewed frequently, but we don't know how this happens.

Ganymede is next out; it is the largest satellite in the solar system, similar to Mercury in size but composed largely of ice. The surface has older, dark, and heavily cratered regions and younger, light regions, which exhibit tectonic features such as parallel ridges and troughs.

Callisto is the outermost of the Galilean satellites. It is also Jupiter's most heavily cratered satellite, with crater densities that suggest the surface is very old. Like Ganymede, Callisto is mostly ice.

In composition, the Galilean moons mimic the solar system itself, with rocky moons close to the center and icy ones further out. Perhaps this reflects a thermal gradient away from a hot young Jupiter when the satellites accreted.

* * *

Saturn is the second largest planet. Like Jupiter it is gaseous, composed mostly of hydrogen and helium, with a small rocky core comprising 25 percent of the planet mass. Unlike Jupiter, however, the atmospheric circulation patterns are obscured by high-altitude haze. Winds as strong as 1,800 kilometers per hour blow eastward along Saturn's equator. These wind velocities decrease markedly towards the poles. Beneath the haze, one can see that Saturn's atmospheric circulation patterns resemble those of Jupiter. Saturn generates almost two and a half times as much heat as it receives from the Sun. How? Probably by gravitational separation of hydrogen and helium. Just as water condensation in Earth's atmosphere causes hot spots, hot spots in Saturn's atmosphere may be caused by helium condensation.

In July of 1610 Galileo saw something that looked like teapot handles around Saturn, but in 1612 they seemed to disappear. In 1655 Christian Huygens suggested that Saturn was ringed by a continuous disk. In 1675 Cassini discovered a break in the disk. A century later Laplace showed that the object around Saturn could not be a continuous disk because gravitational and rotational forces would rip it apart. He proposed a series of narrow, unconnected rings. In 1857 James Clerk Maxwell showed that the rings are composed of individual particles. In 1895 red shift observations showed that the inner rings rotate faster than the outer rings.*

Following the flybys of *Pioneer 11* (1979), *Voyager 1* (1980), and *Voyager 2* (1981), we now have a great deal of information about the rings of Saturn. The ring system extends laterally for 275,000 kilometers, which is three quarters of the distance from the Earth to the Moon. The rings are less than 1 kilometer thick and are composed of snowballs, or ice-coated rocks, ranging from dust to objects no more than a few meters across. The axial and radial structure of the rings is complex. There are seven major radial sections, separated by vacant gaps, and there are spoke-like features, which may result from interactions with Saturn's magnetic field. The density of parti-

---

*The red shift, or Doppler effect, is the decrease or increase in wavelength of sound or light when the source approaches or retreats from the receiver. It causes the familiar change in pitch of whistles or horns from passing trains or cars.

cles, and their composition, appears to vary, with ice being especially prominent in the inner rings.

Saturn has at least 17 moons. Ice, decorated with impact craters and various darker pigments, covers the larger ones except for Titan. On the larger satellites, water volcanism is apparently as common as basalt volcanism on the inner planets. Average densities are typically 1.1 to 1.4 g/cm^3, indicating the dominance of water, ice, and hydrocarbons. Unlike the Galilean satellites, the moons of Saturn are not arranged as a miniature solar system, with the densities decreasing outward. If anything, there seems to be a tendency for densities to increase outward.

Titan is by far the largest of Saturn's moons. Its radius is 2,575 kilometers; its mass is 10^{26} grams; and its density is 1.88 g/cm^3 (compare with the Earth's Moon at 1,738 kilometers, 7.35×10^{25} grams, and 3.34 g/cm^3). Christian Huygens discovered Titan in March of 1655. Three centuries later Gerald Kuiper discovered that Titan has an atmosphere, which is dominated by nitrogen with less than 20 percent methane and argon. This atmosphere is not homogeneous. At the time of the Voyager flybys in 1980 and 1981 it was darker in the north than in the south. It has been likened to a photochemical smog, raining hydrocarbon aerosols such as ethane (C_2H_6), into seas of liquid hydrocarbons. Atmospheric pressure at the surface is 1.5 bars. The temperature at the surface is –179°C.

Beneath the smog and sea of "gasoline," Titan may have a mantle of water ice underlain by a small rocky core.

An interesting question is why Titan has an atmosphere while Jupiter's large satellites, Ganymede and Callisto (each 10^{26}g), are naked. The difference may have been temperature. Perhaps Saturn and its environment were never as warm as Jupiter and its surroundings. The snowflakes that accreted to form Ganymede and Callisto may have been so much warmer than those that formed Titan that they trapped less gas.

* * *

Moving outwards from Saturn to Uranus and Neptune, we find that the satellites become darker and more dense, suggesting increasing proportions of rock to ice and hydrocarbons. Of the 15 known satellites of Uranus, 10 were discovered by *Voyager 2* in 1985 and 1986. The most interesting is Miranda, which looks like a body that was broken apart and reassembled several times but lacked sufficient internal energy to blend the different pieces together into a geologically uniform whole. Neptune is known to have eight satellites, six of which were discovered by *Voyager 2* in 1989. Of these, the largest is Triton, discovered in 1846. Triton's dimensions are radius 1,350 kilometers, mass 2.14×10^{25} grams, and density 2.07 g/cm^3. Its surface temperature is a toasty –235° C. Triton is covered by extensive ice sheets of frozen nitrogen containing a few percent of dissolved carbon monoxide and methane, and separate deposits of frozen carbon dioxide are prominent. Geysers of unknown fluids, perhaps liquid nitrogen mixed with hydrocarbons, are active today. Lavas of ice and frozen ammonia once flowed across impact basins. A tenuous atmosphere of nitrogen and methane has been detected.

Pluto was not visited by the Voyagers, so we know much less about it. Spectroscopic studies from Earth indicate that, like Neptune's moon Triton, the surface of Pluto is dominated by frozen nitrogen with small quantities of dissolved methane and carbon monoxide. Unlike on Triton, where carbon dioxide is prominent, carbon dioxide has yet to be seen on Pluto.

The outer planets and their satellites are strange and wonderful. But one point to bear in mind is that these worlds are *very* far away. Except for Pluto, each is approximately twice as far from the Sun as its inner neighbor:

Jupiter	5.2028 AU
Saturn	9.5388 AU
Uranus	19.1820 AU
Neptune	30.0577 AU
Pluto	39.5177 AU

* * *

Long-haired stars, *aster komates,* or *comets* as we know them, are traditionally the most portentous objects in the solar system. We remember their visits and associate them with calamitous events of human history. All that, of course, is nonsense. Or is it? Why are comets portents of disaster rather than good fortune?

Could it be that, back in the misty shadows of prehistory, our forebearers suffered some unpleasant encounters with them? According to astronomer Fred Whipple, who has studied comets all his life, a comet is simply "a celestial fountain spouting from a large dirty snowball floating through space."[114]

Comets are very different from other objects in the solar system. Whereas asteroids appear as points of light, comets have a nebulous, fuzzy appearance. Every year, about four comets pass through the solar system. Most come in, on hairpin parabolic orbits, from outer space, around the Sun, and back out. Some are tamed by the gravity of the large planets to follow elliptical orbits, returning every 5 to 10 years until they are destroyed. Others take longer to return: Halley's comet, which visited in 1986, will return in 2062, and Comet Kohoutek won't be back for a million years. Of the more than 600 comets whose orbits are known, over 500 have periods exceeding 200 years.

Comets are visible only when they approach within three astronomical units of the Sun. This is within the asteroid belt, between the orbits of Mars (1.5 AU) and Jupiter (5.2 AU). In 1950 Fred Whipple proposed, correctly as it turns out, that a comet has a core of ice and dirt, which sublimates (evaporates without melting) when the comet approaches the Sun. The core, or nucleus, contains a large number of carbon, hydrogen, oxygen, and nitrogen compounds, as well as metals and silicates. As the comet approaches the Sun, the volatile constituents of the core evaporate and are blown away from the Sun by the solar wind. That is why the tails of a comet, which include dust, ionized gas, and plasma, always point away from the Sun. While traveling within three astronomical units of the Sun, the cometary nucleus is surrounded by a spherical region of glowing gas and dust known as the *coma*, which extends out from the nucleus for 10^5 or 10^6 kilometers. The comet's tails may extend as far as 10^8 kilometers.

Whipple's ideas were largely confirmed in 1986 when five spacecraft from Earth made close observations of Comet Halley. The most spectacular observations were those of the European Space Agency's *Giotto* spacecraft, which penetrated the cloud of gas surrounding the nucleus and came within 600 kilometers of its surface. The image Giotto sent back showed a peanut-shaped object, 16 kilometers long, with gas jets spewing out 20 tons of water, dust, carbon oxides, methane, ammonia, and hydrogen cyanide per second. The nucleus was composed of ices, stony materials, and heat-resistant, lightweight organic polymers. Its average density was less than that of water (0.1 to 0.3 g/cm^3), and its reflectivity was extremely low. The nucleus of Comet Halley is the darkest object known in the solar system.

From Aristotle, in the fourth century before Christ, until Tycho Brahe in 1577, comets were thought to originate in Earth's atmosphere. The driving notion was that objects as far away as the Moon and beyond were, by the very nature of the universe, unchanging. Thus, since comets were known to change they must lie within the sphere of Earth. Brahe, the last great pretelescopic astronomer and the person who brought pretelescopic astronomy to the limits of technical precision, searched for a parallax effect* to determine the height of the comet of 1577 above the Earth. Finding no such effect, he suggested that the comets must be at least four times as far from the Earth as the Moon. With this observation, and a similar observation of a supernova in 1572, Brahe tossed the Aristolean idea of a universe constructed of successive, concentric spheres into the dust bin, where it remains today.

Cometary orbits are more eccentric than planetary orbits, and they lie at all angles to the plane of the ecliptic. Some are prograde (in the same direction as Earth), and some are retrograde. Where do comets come from? In 1950 a Dutch astronomer, Jan Oort, studied the orbits of 19 long-term comets and proposed that they come from a huge, spherical cloud around the solar system. The *Oort Cloud,* containing 10^{11} average-sized cometary nuclei, extends more than halfway to the nearest star and contains 10^{25} to 10^{30} grams of materials (compare the Earth's mass at 6×10^{27} grams). In spite of its substantial mass, however, the Oort cloud is so vast that the spacial density of comets is actually less in the Oort Cloud than it is in the inner solar system.

And where did the Oort Cloud itself come from? We don't know. The most commonly held view assumes that the matter

---

*Nearby objects, such as the Moon, are found at different positions in the sky when viewed from different locations; the positions of distant objects, such as stars, do not change.

belongs to the solar nebula. A less common view is that the Oort Cloud was captured from a dense interstellar cloud.

Have comets contributed volatile materials, particularly water and carbon compounds, to Earth? Some scientists have suggested this. One of their arguments is that Earth must have lost a great deal of such material during the hypothetical, core-forming Iron Catastrophe and the ensuing great meltdown.

Other scientists have speculated that comets may also have delivered the seeds of life. This idea, known as *Panspermia* is a radical one, whose principal proponent has been Sir Fred Hoyle, a brilliant astrophysicist. The argument goes as follows: Experiments show that constituents of nucleic acids and proteins can be synthesized by abiotic processes. Simple, lightweight organic molecules are known to exist in comets. Such molecules, irradiated by high-energy electrons, might be transformed into more complex molecules. If life turns out to be cosmic, comets might be the vehicles that deliver it to planetary surfaces. Hoyle and his colleagues argue that the panspermia concept is no less plausible than the spontaneous generation of even a few enzymes on the primitive Earth. Good idea? Bad idea? That is not the question,to ask. Certainly, we should not be like Lawrence Morley's sophisticated critic, who relegated Morley's explanation of seafloor magnetic anomalies to cocktail party conversation. No, the question to ask is whether or not the idea of panspermia can be tested, and the answer is yes. Before long, we will visit a comet and look closely at the materials in its head.

The beauty, mystery and excitement of comets was captured during the recent encounter with Comet Halley by physicist and poet, George Wetherill:

> Among the eucalyptus trees,
> Green leaves dancing in the autumn wind,
> The cold pale watcher of mankind
> Treads his ancient trail again.
>
> Pass swiftly by the angry bull,
> The starry fish and water jar,
> Defy the Sun's consuming flame,
> The archer's bow,
> The scorpion's sting,

The centaur's wrath,
The deadly coil of the hydra—
But then be gone.
Ask not for Harold of Hastings,
You know he is not here;
Nor Attila, vanquished at Chalons,
Edmund, master of Isaac's rules,
Nor Giotto, and the Zealots of Jerusalem.

You must have seen
The ships that rose to greet you.
Next time there will be more.
They'll even mount your haggard head
And ride you into Neptune's night!
Yes, we still are bold.
Though once more we now learn
The message that you bear,
Resonate to your grim tattoo,
The gravest rhythm of our race,
Yet wait with hope your sure return.*

---

*Reprinted with permission from *Year Book 85*, Carnegie Institution of Washington. Copyright by George Wetherill.

Afterword

As we approach the turn of the millennium, the possibilities for increased understanding of the Earth and the solar system are more exciting than ever before. In the decades just ahead our technologies will permit us to reach (or recreate) and examine every environment in the solar system, from the madding heart of the Sun to the vast and empty stillness of the Oort Cloud. We will be there, and we will learn the secrets.

As in decades past, however, we will also use our technologies and our increased understanding of earth and planetary science to better our lot on Earth. We have the means of saving tens of thousands of lives every year by applying our knowledge of Earth to the prediction and mitigation of natural hazards. Just as measles, smallpox, and polio can be eradicated, so, too, most accidental deaths due to volcanic eruptions, earthquakes, floods, avalanches, and great storms can be avoided if existing knowledge is made available and applied. We will be there, and we will share and apply our knowledge.

We will use our increased understanding to become better stewards of the Earth. Knowledge of extraterrestrial space will not alleviate our problems at home. The windswept plains of Mars will never save humans dying on the more hospitable plains of Somalia. The seas of Titan will never fuel our cars, nor will the permafrost of Mars slake the thirst of Los Angeles. But, like the sight of the hangman's noose, the view of Earth from afar has a wonderful concentrating effect on the mind. The problems of Earth have terrestrial solutions, and we will find them.

As I said at the beginning of this book, "Science is demonstrably the most effective technique yet devised for exploring the universe with the human mind." It is also *potentially* the most effective technique for bettering the condition of all who live on Earth. Surely, we will do what needs to be done.

* * *

References

1. Galilei, G. *The Assayer* (1623). Translated in S. Drake and C. D. O'Malley, *The Controversey on the Comets of 1618*. Philadelphia: University of Pennsylvania Press, 1960, p. 184. Reprinted with permission.
2. McKeon, R. (Ed.). *The Basic Works of Aristotle*. New York: Random House, 1941, p. 698.
3. Wigner, E. P. "The Unreasonable Effectiveness of Mathematics in the Natural Sciences." *Communications in Pure and Applied Mathematics*, Vol. 13, 1960, pp. 1–14.
4. Haldane, J. B. S. *Possible Worlds and Other Essays*. London: Chatto and Windus, 1927, p. 286.
5. Hawking, S. *A Brief History of Time*. New York: Bantam Books, 1988, p. 175. Reprinted with permission.
6. Ramo, C. "Black Velvet Skies." *Boston Globe*, June 5, 1989, p. 28.
7. Rubin, V. "The Rotation of Spiral Galaxies." *Science*, Vol. 220, 1983, pp. 1339–1344.
8. Fraunhofer, J. (1817). Quoted in H. Shapley and H. E. Howarth, *A Source Book in Astronomy*. New York: McGraw-Hill, 1929, p. 196.
9. Anders, E., and Ebihara, M. "Solar-System Abundances of the Elements." *Geochimica et Cosmochimica Acta*, Vol. 46, 1982, pp. 2363–2380.
10. CRC Press. *Handbook of Chemistry and Physics* (65th ed.), 1984, p. F–107.
11. Dana, J. D. *Manual of Geology*. New York: Ivison, Blakeman Taylor & Co., 1875, p. 322, Fig. 617.
12. Tschermak, G. *Lehrbuch der Mineralogie*. Vienna: Hölder, 1888, p. 87, Fig. 198.
13. Snow, C. P. *Variety of Men*. New York: Charles Scribner's Sons, 1966, p. 11. Copyright C. P. Snow 1967, reproduced by permission of Curtis Brown Group Ltd., London.
14. Bernard, C. *Introduction à l'Étude de la Médecine Expérimentale* (1865). In J. Barlett, *Familiar Quotations* (16th ed.), J. Kaplan (Ed.), Boston: Little, Brown, 1992, p. 473.

15. Descartes R. *Principles of Philosophy*. In The *Philosophical Works of Descartes*. Translated by E. S. Haldane and G. R. T. Roos, New York: Dover, 1955.
16. Wood, A. *Thomas Young Natural Philosopher 1773–1829*. Cambridge: Cambridge University Press, 1954, p. 166.
17. Röntgen, W. C. "Eine neue Art von Strahlen" (1896). Translated by W. R. Nitske in *The Life of Wilhelm Conrad Röntgen*. Tucson: University of Arizona Press, 1971, App. 3. Reprinted with permission.
18. Meyers, R. A. (Ed.). *Encyclopedia of Modern Physics*. San Diego: Academic Press, 1990, p. 438.
19. Rohrer, H. "The Rise of Local Probe Methods." In H.-J. Güntherodt and R. Wiesendanger (Eds.), *Scanning Tunneling Microscopy I*. Berlin: Springer-Verlag, 1992, p. 14. Reprinted with permission.
20. Mason, B. *Principles of Geochemistry*. New York: John Wiley & Sons, 1966, p. 48.
21. Felix Chayes, 1971, unpublished manuscript.
22. Sorby, H. C. "Fifty Years of Scientific Research." Address delivered to the Sheffield Literary and Philosophical Society, February 2, 1897. In C. H. Summerson, *Sorby On Geology*. Miami: Rosenstiel School of Marine & Atmospheric Science, 1978, p. 217.
23. Snow, C. P. *The Two Cultures and The Scientific Revolution*. New York: Cambridge University Press, 1959, p. 16. Reprinted with the permission of Cambridge University Press.
24. Nicholls, P. (Ed.). *The Science in Science Fiction*. New York: Alfred A. Knopf, 1983, p. 86.
25. Clausius, R. "Ueber die bewegende Kraft der Wärme." *Annalen der Physik und Chemie*, Vol. 79, 1850.
26. N. L. Bowen to J. Gilluly, quoted in H. S., Yoder, "Norman L. Bowen (1887–1956), MIT Class of 1912, First Predoctoral Fellow of the Geophysical Laboratory." *Earth Sciences History*, Vol. 11, 1992, p. 51.
27. Agricola, G. *De Re Metallica* (1556). Translated by H. C. Hoover and L. H. Hoover. New York: Dover Publications, 1950, p. xxx.
28. Based on data from F. Press and R. Siever, *Earth* (4th ed.). New York: W. H. Freeman and Co., 1986, p. 367, Table 14-1.
29. Strutt, R. J. *Proceedings of the Royal Society*, Series A, Vol. 77, 1906, pp. 472–485.
30. Data from B. A., Bolt, *Inside the Earth*. San Francisco: W. H. Freeman and Co., 1982, p. 160, Fig. 7.5.
31. Dana, J. D. *Manual of Geology*. New York: Ivison, Blakeman, Taylor & Co., 1875, p. 108, Fig. 115.
32. Werner, A. G. *Short Classification and Description of the Different Rocks* (1786). Translated by A. M. Ospovat, New York: Hafner Publishing Co., 1971, p. 86, note (n).

33. D'Aubuisson, J. *Traité de Géognosie* (1819). Quoted in A. Geikie, *The Founders of Geology* (2nd ed.). London: Macmillan, 1905, p. 245.
34. Geikie, A. *The Founders of Geology* (2nd ed.). London: MacMillan, 1905, p. 253.
35. Reed, H. H. "Granites and Granites." *Origin of Granite*. Geological Society of America Memoir 28, 1948, 1–19.
36. Hutton, J. A. "Theory of the Earth." *Transactions of the Royal Society of Edinburgh*, Vol.1, 1788, p. 304.
37. Yoder, H. S. "Norman L. Bowen (1887–1956), MIT Class of 1912, First Predoctoral Fellow of the Geophysical Laboratory." *Earth Sciences History*, Vol. 11, 1992, p. 47.
38. Bowen, N. L. "Magmas." *Bulletin of the Geological Society of America*, Vol. 58, 1947, p. 263. Reprinted by permission.
39. Barus, C. "High Temperature Work in Igneous Fusion and Ebullitions, Chiefly in Relation to Pressure." *U.S. Geological Survey Bulletin No. 103*, 1893.
40. Anderson, T., and Flett, J. S. "Report on the Eruption of the Soufrière in St. Vincent, and on a Visit to Montagne Pelée in Martinique." *Philosophical Transactions of the Royal Society of London*, Series A, Vol. 200, 1903, p. 431.
41. Schilling, J-G, Unni, C. K., and Bender, M. L. "The Origin of Chlorine and Bromine in the Oceans." *Nature*, Vol. 273, 1978, pp. 631–636.
42. Walker, D. A. "More Evidence Indicates Link Between El Niños and Seismicity." *EOS*, Vol. 76, No. 4, 1995, pp. 33–36.
43. Gillespie, R. J., Humphreys, D. A., Baird, N. C. and Robinson, E.A. *Chemistry*. Boston: Allyn and Bacon, 1986, p. 479; and Atkins, P. W. *General Chemistry*. New York: Scientific American Books, 1989, p. 360.
44. Holland, H. D. *The Chemistry of the Atmosphere and Oceans*. New York: Wiley-Interscience, 1978, Table 5-1. (Primary sources: for sea water, Pykowicz, R. M., and Kester, D. R. "The Physical Chemistry of Sea Water." *Annual Review of Oceanographic Marine Biology*, Vol. 9, 1971, pp. 11–60; for river water, Gibbs, R. "Water Chemistry of the Amazon River." *Geochimica et Cosmochimica Acta*, Vol. 36, 1972, pp. 1061–1066.)
45. Milliman, J. D. and Meade, R. H. "World-Wide Delivery of River Sediment to the Oceans." *Journal of Geology*, Vol. 91, 1983, pp. 1–21.
46. Blatt, H. *Sedimentary Petrology*. New York: W. H. Freeman and Co., 1982, p. 288.
47. Hutton, J. *Theory of the Earth*. Edinburgh: Cadell, Junior and David, Vol. 1, 1795, pp. 375–376. (Reprinted in 1972 by Verlag von J. Cramer, Lehre).
48. Hutton, J. *Theorie of the Earth*. Edinburgh: Cadell, Junior and David, Vol. 1, 1795, p. 251. (Reprinted by Verlag von J. Cramer, Lehre).

49. Dean, D. *James Hutton and the History of Geology*. Ithaca, New York: Cornell University Press, 1992, p. 141.
50. Chambers, Robert, *Domestic Annals of Scotland*. Edinburgh: W & R Chambers, Vol. 2, 1859, p. 136.
51. Edmond, J. and Von Damm, K. "Hot Springs on the Ocean Floor." *Scientific American*, Vol. 248, No. 4, 1983, p. 86. Reprinted with permission.
52. Macdonald, K. C. and Luyendyk, B. P. "The Crest of the East Pacific Rise." *Scientific American*, Vol. 244, No. 5, 1981, pp. 104–105. Reprinted by permission.
53. Hutton, J. *Theory of the Earth*. Edinburgh: Cadell, Junior and David, Vol. ii, 1795, p. 562. (Reprinted 1972 by Verlag Von J, Cramer, Lehre)
54. Bartlett, J. *Familiar Quotations* (16th ed.). Boston: Little, Brown and Co., 1992, p. 555.
55. Le Conte, J. *Elements of Geology*. New York: Appleton, 1895, p. 180, Fig. 157.
56. Playfair, J. *Collected Works of John Playfair Esq.*, Vol. 4, 1822, pp. 33–118.
57. Ussher, J. A. *Annales Veteris Testimenti, a prima mundi origine deducti.*(1650). Passage translated by L. C. Bruno in H. S. Yoder "Timetable of Petrology." *Journal of Geological Education*, Vol. 41, 1993, pp. 447–489.
58. Lyell, C. *Principles of Geology*. London: John Murray, Vol. 3, 1833, p. 385.
59. Berner, E. K., and Berner, R. A. *The Global Water Cycle*. Englewood Cliffs, N.J.: Prentice-Hall, 1987, Table 8.3.
60. Residence times are from H. D. Holland, *The Chemistry of the Atmosphere and Oceans*. New York: Wiley-Interscience, 1978, Tables 5-1 and 5-2.
61. Chamberlain, T. C. *Journal of Geology*, 1899.
62. Rutherford, E. quoted in D. E. Wilson, *Rutherford Simple Genius*. Cambridge, Mass.: MIT Press, 1983, p. 206. Reprinted with permission.
63. Letter from Lord Kelvin to T. E. Young, June 9, 1906. Quoted in C. Smith and M. N. Wise, *Energy and Empire: A Biographical Study of Lord Kelvin*. Cambridge: Cambridge University Press, 1989, p. 611. Reprinted with the permission of Cambridge University Press.
64. Le Conte, J. *Elements of Geology*. New York: Appleton, 1895, p. 253, Fig. 223.
65. Richards, J. A., Sears, F. W., Wehr, M. R., and Zemansky, M. W. *Modern University Physics*. Reading, Mass.: Addison-Wesley, 1960, p. 88.
66. Newton, I. *Principia*. Lib. III, 1685–7. Propositio X, Theorema X. Translated by F. Cajori. Berkeley: University of California Press, 1934, p. 418.
67. "Donnaeha Ruadh (Red-Haired Duncan)." *A Bhan Lunnainneach Bhuidhe*. Translated in D. Howse, *Nevil Maskelyne, The Seaman's Astronomer*. Cambridge: Cambridge University Press, 1989, pp. 137–138. Reprinted with the permission of Cambridge University Press.

68. Bolt, B. A. *Inside The Earth*. San Francisco: W. H. Freeman, 1982, p. 8.
69. Temple, R. *The Genius of China*. New York: Simon and Schuster, 1986. p. 163. Reprinted with permission.
70. Oldham, R. D. "The Constitution of the Interior of the Earth as Revealed by Earthquakes." *Geological Society of London Quarterly Journal*, Vol. 62, 1906, p.456–475, Fig. 2.
71. Bolt, B. A. *Inside The Earth*. San Francisco: W. H. Freeman, 1982, Table 4.1.
72. Anderson, D. L. "Viewpoint: The Future." *Science*, Vol. 267, 1995, p. 1618.
73. Jeanloz, R., and Lay, T. "The Core-Mantle Boundary." *Scientific American*, Vol. 268, No. 5, 1993, pp. 48–55.
74. Winfree, A. T. *The Timing of Biological Clocks*. New York: Scientific American Books, 1987, p. 47.
75. Wegener, A. *Die Enstehung der Kontinente und Ozeane* (1929). Translated by J. Biram. Mineola, N.Y.: Dover Publications, 1966, p. 1.
76. Simpson, G. G. *Concession to the Improbable*. New Haven: Yale University Press, 1978, pp. 272–273. Reprinted with permission.
77. Compton, A. H. *American State Papers*, Misc., I, p. 140.
78. Jeffreys, H. *The Earth* (3rd ed.). Cambridge: Cambridge University Press, 1952, p. 348. Reprinted with permission of Cambridge University Press.
79. Bullard, E. "The Emergence of Plate Tectonics: A Personal View." *Annual Review of Earth and Planetary Sciences*, Vol. 3, 1975, p. 8–10. Reproduced, with permission. Copyright 1975 by Annual Reviews, Inc.
80. Judson, S., Deffeyes, K. S., and Hargraves, R. B. *Physical Geology*. Englewood Cliffs, N.J.: Prentice-Hall, 1976, p. 205. Reprinted with permission.
81. Hess, H. H. "Drowned Ancient Islands of the Pacific Basin." *American Journal of Science*, Vol. 244, 1946, pp. 772–791.
82. Shagam, R. (Ed.). *Studies in Earth and Space Sciences*. The Geological Society of America Memoir 132, 1972, pp. xiv–xv.
83. From an address given by Louis Pasteur at the University of Lille, December 7, 1854.
84. Morley, L. "Early Work Leading to the Explanation of the Banded Geomagnetic Imprinting of the Ocean Floor. *EOS*, Vol. 67, No. 36, 1986, p. 666. Copyright by the American Geophysical Union. Reprinted with permission.
85. Meadows, D. H., Meadows, D. L., Randers, J., and Behrens, W. W. *The Limits to Growth*. New York: Universe Books, 1972.
86. Malenbaum, W. *World Demand For Raw Materials In 1985 And 2000*. New York: McGraw-Hill, 1978, p. 2. Reprinted with permission.

87. Skinner, B. J., and Porter, S. C. *Physical Geology*. New York: John Wiley & Sons, 1987, p. 593.
88. Data converted from K. B. Krauskopf, *Radioactive Waste Disposal and Geology*. New York: Chapman and Hall, 1988, Table 2.1.
89. U.S. Geological Survey. *The High Plains Aquifer*. Water Fact Sheet. Reston, VA., 1983.
90. World Resources Institute. *World Resources 1986*. New York: Basic Books, 1986; and Forkasiewicz, J. and Margat, J. *Tableau Mondial de Données nationales d'Économie de l'Eau. Resources et Utilisations*. Département Hydrologéologie, 79 SGN 784 HYD, Orleans, France, 1986. Reported in A. T. McDonald and D. Kay, *Water Resources Issues & Strategies*. New York: John Wiley & Sons), 1988, Table 4.1.
91. Hammons, T. J. "Tidal Power Potential." *IEEE Power Engineering Review*, Vol. 13, No. 3, 1993, pp. 3–17.
92. Cañada Guerrero, F. "Las Corrientes del Estrecho de Gibraltar Como una Importante Fuente de Energía. Ampliación del Esquema Sobre un Proyecto para su Aprovechamiento. Nuevas Perspectivas para Instalaciones Maremotrices." *Boletín Geológico y Minero*, Vol. XCII-II, 1981, pp. 141–150.
93. Austin, G. T. *Gemstones 1993*. U.S. Department of the Interior, Bureau of Mines Annual Report, 1993, p. 25–26, Table 13.
94. Campbell, J. *Trav S Afr.* (2nd ed.), Vol. xxvii, 1815, p. 349. In *OED*, Vol. VI, p. 854.
95. Bruton, E. *Diamonds*. Radnor, Pa.: Chilton Book Co., 1970, p. 30.
96. McGetchin, T. R., and Ulrich, C. W. "Xenoliths in Maars and Diatremes with Inferences for the Moon, Mars and Venus." *Journal of Geophysical Research*, Vol. 78, No. 11, 1973, pp 1833–1853.
97. Bridgeman, P. W. "Synthetic Diamonds." *Scientific American*, Vol. 193, No. 11, 1955, p. 44.
98. Bundy, F. P., Bassett, W. A., Weathers, M. S., Hemley, R. J., Mao, H-K., and Goncharov, A. L. "The Pressure-Temperature Phase and Transformation Diagram for Carbon; Updated to 1994." *Carbon*, (In Press).
99. Beatty, J. K., and A. Chaiken (Eds.). *The New Solar System* (3rd ed.). Cambridge, Ma.: Cambridge University Press & Sky Publishing Corporation, 1990, p. 298.
100. Adams, M. M. *William Ockham*. South Bend, Indiana: University of Notre Dame Press), 1987, p. 156.
101. Wood, J. A., Dickey, J. S., Marvin, U. B., and Powell, B. N. "Lunar Anorthosites." *Science*, Vol. 167, 1970, pp. 602–604.
102. Robinson, L. J. (Ed.). "Ski Mercury!" *Sky & Telescope*, Vol. 88, No. 4, 1994, p. 12.
103. Lomonosov, M. V. "The Appearance of Venus on the Sun, Observed in the St. Petersburg Academy of Sciences on the 26th Day of May of the

year 1761." Quoted in B. N. Menshutkin, *Russia's Lomonosov: Chemist, Courtier, Physicist, Poet.*. Translated by J. E. Thal and E. J. Webster. Princeton, N. J.: Princeton University Press, 1952, p. 147.

104. Phillips, R. J., Kaula, W. M., McGill, G. E., and Malin, M. C. "Tectonics and Evolution of Venus." *Science*, Vol. 212, No. 4497, 1981, pp. 879–887, Fig. 2.
105. Van Allen, J. A. "Magnetospheres, Cosmic Rays, and the Interplanetary Medium." In J. K. Beatty and A. Chaikin (Eds.), *The New Solar System* (3rd ed.). Cambridge, Ma.: Cambridge University Press & Sky Publishing Corporation, 1990, pp. 29–40.
106. Phillips, R. J., and Hansen, V. L. "Tectonic and Magmatic Evolution of Venus." *Annual Review of Earth and Planetary Sciences*, Vol. 22, 1994, p. 649.
107. Kepler, J. *Astronomia nova* (1609). Translated in O. Gingerich, *Laboratory Exercises in Astronomy —The Orbit of Mars*. Cambridge, Ma.: Sky Publishing Corporation, 1983.
108. Mustard, J. F., and Sunshine, J. M. "Seeing Through the Dust: Martian Crustal Heterogeneity and Links to the SNC Meteorites." *Science*, Vol. 267, 1995, pp. 1623–1626.
109. Shoemaker, E. M., and Shoemaker, C. S. "The Collision of Solid Bodies." In J. K. Beatty and A. Chaiken (Eds.), *The New Solar System* (3rd ed.). Cambridge, Ma.: Cambridge University Press & Sky Publishing Corporation, 1990, p. 260.
110. Shoemaker, E. M., and Shoemaker, C. S. "The Collision of Solid Bodies." In J. K. Beatty, and A. Chaiken (Eds.), *The New Solar System* (3rd ed.). Cambridge, Ma.: Cambridge University Press & Sky Publishing Corporation, 1990, p. 261, Table 2.
111. Shoemaker, E. M., and Shoemaker, C. S. "The Collision of Small Bodies." In J. K. Beatty and A. Chaiken (Eds.), *The New Solar System* (3rd ed.). Cambridge, Ma.: Cambridge Universiy Press & Sky Publishing Corporation, 1990, p.274. Reprinted with permission.
112. Milne, A. A. *The House at Pooh Corner*. New York: E. P. Dutton & Co., 1928, p.80.
113. Galilei, G. *Sidereus Nuncius* (1610). Translated by A. Van Helden, Chicago: The University of Chicago Press, 1989, p. 64. Reprinted with permission.
114. Whipple, F. L. "The Nature of Comets." *Scientific American*, Vol. 230, No. 2, 1974, p. 49.

Index

Index